Relativity, Gravitation and GPS

Filling the Voids

Second Edition

Clive A. Redwood

Send feedback to caredwoodgm@gmail.com

To Ivy, my wife and cherished partner, whose love, care, support, and encouragement has made my life so wonderful and this work a meaningful pleasure.

Relativity, Gravitation and GPS

CONTENTS

Preface to First Edition

Probably, like me, you have too often been frustrated by the logical discontinuities in scientific and technical writings. These gaps are frequently preceded by announcements such as 'it can easily be shown that', 'after some manipulation' or 'elsewhere in the literature' and so on. Quite often there is simply an axiomatic assertion of a concept or of a procedure. One is then left with the options of either accepting the logical consistency of the presentation on the basis of faith (a proposition that is fraught with contradictions) or trying to bridge these gaps by means of one's own endeavours.

Of course, one supposes that there may be perfectly good reasons for these discontinuities. One such reason may be that the presentations were intended for the initiated. One may also rationalize that the reason for the gaps lies in the reduction of the cost of publication. However, regardless of the actual justifications, rationally following the presentation is inhibited. This leaves the uninitiated reader somewhat out of sorts.

This work is substantially based on Einstein's *The Foundation of the General Theory of Relativity* and Neil Ashby's *Relativity in the Global Positioning System*. Both of these works are quite compact and seem to have been written for the initiated. The sub-title of this work – *Filling the Voids* - announces my intention of not leaving any logical gaps in the presentation, excepting for those resolved by basic algebraic manipulations and calculus. Hopefully, having read this work, the reader will agree that this objective has been met.

C. A. R

Preface to Second Edition

This edition improves the presentation of the first edition mainly in two sections. The section on the anomalous precession of Mercury's orbit has been completely revamped. A more direct approach is here employed without alluding to a heuristic as was done before. Secondly, the determination of the local time in the section on the Global Positioning System (GPS) has been improved to reveal more clearly the stages through which the final estimate is achieved.

In addition, two sections have been added. In the main text, a section on the behaviour of light in weak gravity has been added. Both the deflection and the delay of light in the field are presented. Incorporation of the consideration of this delay would improve the accuracies of the GPS estimates. This may appear in the next generation of global navigation systems.

In the Appendix, along with the discussion on the orbital ellipse from the first edition, an alternative approach to the derivation of the field equations of gravity is now presented. This derivation proceeds directly from a variational approach. The importance of this approach is that it is common to approaches in other areas of physics outside of relativity and gravitation, notably in particle physics. Furthermore, conserved quantities may thereby be indicated.

C. A. R

Acknowledgements

I am grateful for the advice that I received from my former teacher – Professor Michael McMorris – in the preparation of this work. The introduction to modern physics that I received as his student and his more recent guidance have made possible this work and has much improved its presentation, respectively.

My good friend and colleague – Milton Bennett – has kindly edited several versions of this work. Without his attention to detail, tirelessness and meticulousness in divining and listing the erroneous commissions, devilish omissions, and subtle ambiguities, this work would not be as readable as I hope it is.

C. A. R

Introduction

The science of relativity and gravitation, essentially established in Einstein's *The Foundation of the General Theory of Relativity (FGTR)*, has long been ensconced in the unpopular heights of high science and cosmological ruminations all, more or less, irrelevant to the mundane world. The presentations on the general theory of relativity (GTR) and on its applications in astrophysics were therefore mostly crafted for scientific specialists. Now, the significance of GTR in the technology of the mundane and peripatetic Global Positioning System (GPS) has made its appreciation necessary even for ordinary persons. Yet recently, the popular and even the scientific accounts of GTR tend to foster a complicating myth: 'the fabric of spacetime'.

As appliances with embedded GPS technology abound, it is imperative that this technology be popularly appreciated. It is important that we understand the things that we use and not just how to use them. This work firstly attempts to increase the accessibility to GTR, not by glossing over its necessary complexity, but by filling in the gaps usually present in publications for specialists and by avoiding too abstract notation. Myths are definitely avoided. Subsequently, the application of the principles of GTR in the technology of GPS is presented.

The prerequisites for following the material are equivalent to a background in: calculus including partial differentiation, integration over lines and volumes and variations; linear algebra including quadratic forms and the differentiation of determinants; and physics including Newtonian theory of gravity and the electromagnetism of Maxwell.

In this work, we have endeavoured to emphasize the revolutionary nature of the science of relativity. Recall that Minkowski had declared:

> "Henceforth, space by itself, and time by itself, are doomed to fade away into mere shadows, and only a kind of union of the two will preserve an independent reality."[1]

However, for Minkowski, light had a fixed speed *in vacuo* regardless of the coordinate system (*chart*) that was used. Furthermore, he is not remembered for considering the revolutionizing possibility that charts may be accelerated. Einstein was enormously inspired by Minkowski's brilliant and revolutionary achievement of recognizing in the Lorentzian contraction of length and dilation of duration a paradigm-changing quadratic relationship of the coordinate intervals that yields the locally measured distance or duration, but he remained unbounded by Minkowski's limitations.

The Minkowskian relationship primarily concerns measurements of lengths and durations in two charts moving, relatively to one another, with constant velocity in a region without gravity. In other words, the Special Theory of Relativity (STR) is primarily concerned with the metrical relationship of inertial charts. Einstein's GTR is concerned with charts moving arbitrarily with respect to one another, even in the presence of gravity. In general, relativity is concerned *only* with the expression of physical laws *with respect to coordinate charts*, beginning with the laws of mensuration. The Principle of Relativity requires *only* that physical laws hold good in various charts, that is, that these laws are *covariant* with respect to transformations from one chart to another chart. In 1915 in his *FGTR*, Einstein boldly declared:

'That this requirement of general covariance... takes away from space and time the last remnant of physical objectivity...'[2]

In a much later publication of 1952, *Relativity and the Problem of Space*, Einstein made it clear that:

'Space-time does not claim existence on its own, but only as a structural quality of the field.'[3]

Thus Einstein cast 'space' and 'time' away from physical *science*. Furthermore, he reduces 'spacetime' from being an entity to merely being an attribute of an entity. How does a quality become a fabric? Such a transformation seems purely ideological.

In our discussion on GTR, use is made of the adjectives *translational* and *transitional* to refer to the 'space' and the 'time' coordinates, respectively. The main virtue of these terms is their operational aspect. This is an attempt in terminology to overtake a bad ideology that is being immersed in habit.

Unfortunately, this ideology has already created a *monstrosity* in the 'spacetime fabric' that rears its invisible head over the minds of physicists. This 'spacetime fabric' is an *undetectable* (and thereby hypothetical) continuum that , it is alleged, not only provides the room for phenomena, as did Newton's Absolute Space, but also is acted upon by physical objects and energetic processes and then itself reacts upon things to cause them to move. For example, John Wheeler is said to have stated:

'Spacetime tells matter how to move; matter tells spacetime how to curve.'[4]

Now, to his scientific credit, Newton having paid philosophical homage to Absolute Space and Absolute Time promptly leaves them at the outset of his science: in fact, further on, he only deals with displacement (as relative space) and 'common time' - the measure of duration by means of motion.[5] However, for some modern physicists, the phenomena of gravity are only to be explained by the 'spacetime fabric': massive objects curve 'spacetime' and consequently objects move 'down' these curves. As in popular publications, so it is in academic texts, and even in scientific papers.

Aside, from everything else, the *circular* nature of this explanation escapes its learned proposers/defenders. They seem to forget that, in the first place, gravity was theorized to explain why objects move *down* (slopes, for example). They, somehow, do not then think it necessary to explain why any movement should result from this wondrous curve formation *of* 'spacetime' and why it is that the resulting movement is necessarily 'downwards' on the 'spacetime fabric'. In fact, they seem to be in need of another potential. So in a real sense, they are still at the original problem that Newton's theory of gravitation was proposed to explain.

Their *retrogressive* error lies in maintaining that coordinate systems really do describe 'spacetime'. The fact that over the same region, different coordinate charts may be simultaneously applied never revealed to them the clue that the metrical properties, that characterize a chart, belong to the chart itself and not to any 'spacetime' of the region.

By means of his revolutionizing recognition of the local *observational* equivalence of a stationary chart in a gravitational field and the acceleration of a chart in a region without a gravitational field - his Equivalence Principle - Einstein was able to extend relativity theory to include gravitational influences[6]. Yet, here he still was primarily concerned with relativity and charts. Gravity was recognized as effective a condition in the construction of charts as is acceleration[7] and so had to be included in the study of relativity.

In return, the science of relativity allowed gravity to be analyzed by the methods that originally had been developed for understanding relativity.[8] Uniquely in GTR, relativity accommodates accelerated frames of reference. On the other hand, since the *dynamical strength (force per unit mass)* of the gravitational field is purely a *kinematical* quantity – acceleration – then the *dynamics* of gravity revolves, so to speak, around kinematical quantities related, by virtue of the Principle of Equivalence, to the metrical properties of charts revealed by GTR.

However, the theory of gravitation departs from relativity considerations by way of the Einstein Field Equations (EFE) and the geodesic equation that describe the *induction* of the energy components of gravity by the energy components of material processes and the unrestricted motion of matter in induced gravitational fields, respectively. Yet Einstein's theory of gravitation revealed the internal dynamical structure of gravity as a field that, in its representation *inescapably* relative to a chart, is comprised of translational and transitional coordinate derivatives of the metrically relevant parameters of relativity theory. These parameters, collectively termed the *metric*, constitute the umbilical cord of the theory of gravitation mothered by the theory of relativity.

Now charts are factitious as Einstein called them.[9] As such they cannot be directly acted upon by physical forces. So it may be this strange effect of a physical (energetic) gravity on the metrical (informational) properties of an observer's chart that seems to have given rise to the concept of the supposedly physically real and extremely significant, but definitely *ghostly*, 'Spacetime Fabric'. Conversely, coordinate charts cannot have real physical effects as is implied by the necessity of the general covariance of the laws of physics and the requirement of their generally covariant formulation. On the whole, the matter may be stated in this way: gravity, induced by the energy of material processes, *energetically* affects material processes and objects and *kinematically* influences charts, the latter by means *only* of its (observational/informational) equivalence to chart acceleration. GTR deals directly with the kinematical and gravitational influences on the metrical (informational) properties of charts.

It is rather regrettable that since the great revolutionary scientists have thrown out of science the dubitable continua - Absolute Space and Absolute Time (Newton), the unmoving and undetectable ether (Einstein in STR), 'time' (Einstein in STR[10]), 'space' (Minkowski in STR), and 'spacetime' (Einstein in GTR) - that some modern physicists seem to have gone *back* to an *allegedly existent and directly undetectable continuum*; 'the spacetime fabric'. This time around the mysterious continuum is also *movable* and allegedly displays *dynamical interactions* with matter.

We may do away with this cumbrous, retrogressive, purely hypothetical concept of the 'spacetime fabric' simply by recognizing that metrical properties do not apply to 'spacetime' but merely to chart transformations (information processing) in order to ensure the invariance of the *proper* results of measurement. That is, the metrical properties of a chart directly express the relationship of general covariance of *the law of the invariance of proper lengths and durations.*[11] (Proper length and duration are the results of measurements performed with respect to an observer's chart wherein the subject, object, and instrument of measurement are at rest and close relatively to one another.)

The invariance of proper length and duration is, in a profound sense, the primary law of physics. The measurements of length and duration when performed in an arbitrary chart result, at first, in *coordinate intervals* that generally require *subsequent* information processing - *chart transformation* - to arrive at the invariant *proper* magnitudes. This is the general covariance of 4-D mensuration. Without such post-measurement metrical transformation, or at least its consideration, coordinate determinations are liable to be *illusory*.

Furthermore, it must not be forgotten that the expressions of curvature in the EFE relate directly to the gravitational field and not to a 'spacetime fabric'. As an independently objective field, gravity has a structure – components, extension, and so on – and forms an energetic continuum. Its analysis is subject to methods similar to those used in the study of other continua including the use of concepts such as continuity, differentiability, and curvature. What is being expressed in the EFE is not 'the bending of spacetime', but the movement and state of *existent* gravitational fields. The metrically-related parameters are dynamically fundamental, but not as descriptors of 'spacetime', but of the gravitational field. Gravity is real: energetic and therefore detectable. 'Spacetime' simply is not.

In the discussion on GPS, the author surrenders to the lexical imperative of technical jargon, so the 'bad' words are used.

As to what are really referred to by the use of the terms 'space', 'time', and 'spacetime', for the author's opinions, please see his poem *The Illusions of Space, Time, and Spacetime* that follows the main text.

Part I

Relativity, Gravitation and GPS

Relativity, Gravitation and GPS

Part 1

A. Initial Assumptions of the General Theory of Relativity and Gravitation

1. 'We assume the possibility of verifying "simultaneity" for events immediately proximate in space or – to speak more precisely – for immediate proximity or coincidence in space-time, without giving a definition of this fundamental concept.'[12]
2. 'Moreover, the results of our measuring are nothing but verifications of such meetings of our measuring instruments with other material points, coincidences between the hands of a clock dial, and observed point-events happening at the same place at the same time.'[2]
3. All coordinate charts are 'equally suitable for the description of nature.'[2] This is the Principle of Relativity.
4. 'The general laws of nature are expressed by equations which hold good for all systems of coordinates, that is, are covariant with respect to any substitutions whatever (generally covariant).'[2] This is the Relativity Postulate.
5. 'That this requirement of general covariance, which takes away from space and time the last remnant of physical objectivity, is a natural one… '[2]
6. A stationary chart in a gravitational field and acceleration of a chart in the absence of gravity are *locally* indistinguishable: the Equivalence Principle.
7. In any infinitesimal region, a Minkowskian chart of coordinates may be established.[13]

B. Generalization of the Metric of the Special Theory of Relativity

Consider the finite linear element of the Special Theory of Relativity (STR):

$$s^2 = (x^1)^2 + (x^2)^2 + (x^3)^2 + (x^4)^2 \tag{1a}$$

where: $x^1 = x \quad x^2 = y \quad x^3 = z \quad x^4 = ict \quad i = \sqrt{-1}$

and c is the speed of light. The linear element s is a scalar – distance or duration - that is *measured locally* and is *invariant* under transformation. The x^μ for $\mu = 1, 2, 3, 4$ are coordinates of a 4-D vector with respect to a given chart. There is no acceleration of this chart and gravity is absent. (When we speak of the velocity or acceleration of a chart it is always with respect to the site or chart of the local measurement of the linear element.)

Minkowski's equation (1a) of the linear element subsumes pre-Minkowskian mensuration including: the Euclidean/Pythagorean metric, Galilean relativity and Lorentzian transformations and is STR's point of departure to GTR.

Rewriting equation (1a):

$$s^2 = \sum_{\mu=1}^{4} (x^\mu)^2$$

where the x^μ are components of a 4-D vector $\bar{\mathbf{x}}$.

Consider a change of basis or, to rephrase, a linear and homogenous transformation to a new coordinate chart denoted by the prime. The inverse of this transformation, obtained by applying Leibniz' chain rule, yields:

$$x^\mu = \sum_{\tau=1}^{4} \frac{\partial x^\mu}{\partial x'^\tau} x'^\tau \qquad \mu = 1, 2, 3, 4$$

So expressing x^μ most generally, we may write:

$$s^2 = \sum_{\mu=1}^{4} \left[\sum_{\tau=1}^{4} \left(\frac{\partial x^\mu}{x'^\tau} x'^\tau \right) \sum_{\sigma=1}^{4} \left(\frac{\partial x^\mu}{\partial x'^\sigma} x'^\sigma \right) \right]$$

As the summations occur over independent indices, we can rewrite this as:

$$s^2 = \sum_{\tau,\sigma=1}^{4} \left[\sum_{\mu=1}^{4} \frac{\partial x^\mu}{\partial x'^\tau} \frac{\partial x^\mu}{\partial x'^\sigma} \right] x'^\tau x'^\sigma$$

However, according to the assumption of paragraph 7 of § A, equation (1a), in general, holds only in an infinitesimal region. So in generalizing (1a), let us then only consider infinitesimal linear elements and vectors:

$$ds^2 = \sum_{\tau,\sigma=1}^{4} \left[\sum_{\mu=1}^{4} \frac{\partial x^\mu}{\partial x'^\tau} \frac{\partial x^\mu}{\partial x'^\sigma} \right] dx'^\tau dx'^\sigma$$

For each pair of values of τ and σ, let:

$$g_{\tau\sigma} = \left[\sum_{\mu=1}^{4} \frac{\partial x^\mu}{\partial x'^\tau} \frac{\partial x^\mu}{\partial x'^\sigma} \right] \qquad \tau, \sigma = 1, 2, 3, 4 \qquad (2)$$

These are sixteen equations. Summing over all permutations of the indices - τ and σ - each independently taken over the set {1, 2, 3, 4}, we have:

$$ds^2 = \sum_{\tau,\sigma=1}^{4} g_{\tau\sigma} dx'^{\tau} dx'^{\sigma} \quad (3)$$

Equations (3) and (2) generalize equation (1a).[14] The sixteen quantities - $g_{\tau\sigma}$ - constitute the *metric*.

Equation (1a) applies to a chart that is unaccelerated in the absence of gravity. Here the $g_{\tau\sigma}$ are constants. In the general case, represented in equation (3), where the chart may be accelerated, the $g_{\tau\sigma}$ are functions of the coordinates. From the Equivalence Principle, it follows that if the chart is stationary in a gravitational field then the $g_{\tau\sigma}$ are related to this gravitational field. We shall consider that *generally* the $g_{\tau\sigma}$ are related to the gravitational field described relative to a chart.

A common error is to regard the metric as being the expression of the '*metrical properties of spacetime*', possibly because it seems that *given* dx'^{τ} and dx'^{σ}, the $g_{\tau\sigma}$ *determine* ds. However, ds is a *proper* measurand, and results either from laying a measuring rod relatively at rest and alongside an object, or from measuring a duration by means of a clock at rest relatively, and proximate, to the observer. Although such a *natural* measurand as ds remains unchanged across all charts of the region, there may be several charts each with its own metric - $g_{\mu\nu}$ - and its own set of *coordinate* intervals – dx^{μ} and dx^{ν} – that, as a group, may be unique for each chart. As such, it is clear that the metric actually mediates the *correspondence* between the *set of coordinate intervals of a chart* and the *invariant quantity* - ds.[15] Therefore for any given ds, a metric and its associated set of corresponding coordinate intervals both belong to a particular chart and not to 'spacetime'.

For $x^4 = ct$ and with infinitesimal intervals, equation (1a) becomes:

$$ds^2 = (dx^1)^2 + (dx^2)^2 + (dx^3)^2 - (dx^4)^2 \quad (1b)$$

In this form, the metric of the Minkowskian chart of the STR may be specified as:

$$g_{\tau\sigma} = 1 \qquad : \tau = \sigma = 1, 2, 3$$
$$g_{\tau\sigma} = 0 \qquad : \tau \neq \sigma$$

and $g_{44} = -1$

Here the linear element has a metric of *signature* $+++-$ and, when positive, is said to be 'space-like'. This metric characterises an unaccelerated chart of a region with zero gravitational strength.

Alternatively, and without compromise, equation (1b) may be written as:

$$ds^2 = (cd\tau)^2 = -(dx^1)^2 - (dx^2)^2 - (dx^3)^2 + (dx^4)^2 \quad (1c)$$

Here the linear element has signature $---+$ and, when positive, is said to be 'time-like'. The quantity $d\tau$ is the observer's proper translational interval (in seconds).

C. Tensor Mathematics Basics

In order to be generally covariant, the laws of nature must be expressed in mathematical forms that are independent of any particular system of coordinates. However, these expressions must be readily translatable to any valid system of coordinates. The generally covariant law must therefore be expressed in a mathematical form that, though essentially exclusive of any coordinate system, has qualities essential to all coordinate systems.

The common essential qualities of all coordinate systems are expressed in certain properties as exhibited in their transformations both within each system of coordinates and between different systems of coordinates. Essentially, all such transformations must exhibit linearity (consistency in scaling) and homogeneity (no constant term introduced by the transformation).

The mathematics of tensor analysis provides just such a platform for generally covariant expressions. The mathematical entity called a *tensor is a component of a linear homogenous transformation*. In using tensors, no particular coordinate system is implied, yet all valid systems of coordinates may be substituted into the covariant expression.

Tensors are classified, in the first place, by the type of transformation to which they are subject. Secondly, they are ranked according to the exponent that raises the dimension of the domain of their application to the number of their components. The rank is directly expressed in the number of indices that are associated with the tensor and the dimension of the tensor's domain is given by the number of elements in the set over which each index is taken.

C1. Tensors of the First Rank

C1.1. The Contravariant 4-Vector

A contravariant tensor of the first rank is the equivalent of a vector. A contravariant transformation is merely a change of coordinate chart. Such a transformation leaves the magnitude and orientation of the vector unchanged.

Transforming a 4-vector (a tensor with four components) to a new chart by means of the chain rule we get:

$$dx'^{\sigma} = \sum_{\nu=1}^{4} \frac{\partial x'^{\sigma}}{\partial x^{\nu}} dx^{\nu}$$

where the dx^{ν} are components of the vector $d\bar{\mathbf{x}}$ and dx'^{σ} are the components of the vector $d\bar{\mathbf{x}}'$.

More generally, the equation:

$$A'^{\sigma} = \sum_{\nu=1}^{4} \frac{\partial x'^{\sigma}}{\partial x^{\nu}} A^{\nu}$$

describes a contravariant transformation. (Note that the superscript denotes contravariant quantities.) The four quantities A^{ν} are objects of a contravariant linear homogenous transformation. As components of a linear homogenous transformation they are tensors.

The linear and homogenous character of the transformation implies:

$$A^{\nu} \pm B^{\nu}$$

are also contravariant 4-quantities, if the B^{ν} are also.

C1.2. The Covariant 4-Vector

If, for an arbitrary contravariant 4-vector B^{ν}, the following holds:

$$\sum_{\nu=1}^{4} A_{\nu} B^{\nu} = \text{invariant (scalar)}$$

then A_{ν} is a tensor. The proof is as follows. Consider;

$$\sum_{\sigma=1}^{4} A'_{\sigma} B'^{\sigma} = \sum_{\nu=1}^{4} A_{\nu} B^{\nu}$$

Here B'^{σ} is the contravariant transform of B^{ν}. It follows that:

$$B^{\nu} = \sum_{\sigma=1}^{4} \frac{\partial x^{\nu}}{\partial x'^{\sigma}} B'^{\sigma}$$

is the inverse transformation. So we may write:

$$\sum_{\sigma=1}^{4} A'_{\sigma} B'^{\sigma} = \sum_{\nu=1}^{4} \left[A_{\nu} \sum_{\sigma=1}^{4} \frac{\partial x^{\nu}}{\partial x'^{\sigma}} B'^{\sigma} \right] = \sum_{\sigma=1}^{4} \left[\sum_{\nu=1}^{4} A_{\nu} \frac{\partial x^{\nu}}{\partial x'^{\sigma}} \right] B'^{\sigma}$$

For arbitrary B'^{σ}, this implies:

$$A'_{\sigma} = \sum_{\nu=1}^{4} \frac{\partial x^{\nu}}{\partial x'^{\sigma}} A_{\nu}$$

This is a linear and homogenous transformation. Therefore A_{ν} is a tensor. The transformation defined by the equation above is of the *covariant* type. Therefore, A_{ν} is a covariant 4-vector tensor.

Index Notation

Note that in the transformations the summation occurs over the indices that appear twice in the same term. With that in mind, we may discard the summation symbol. This convention is called the Einstein Notation. For example;

$$A'_\sigma = \frac{\partial x^\nu}{\partial x'^\sigma} A_\nu$$

implies:

$$A'_\sigma = \sum_{\nu=1}^{4} \frac{\partial x^\nu}{\partial x'^\sigma} A_\nu$$

The following are the *rules* of the manipulation of indices:

The "denomination of summation indices is immaterial"[16] as they disappear in the summation. *Therefore, in an expression, notational change of summation indices may be independently applied uniformly to each term.* These indices on substitution yield four (the dimension of the domain of application) terms.

On the other hand, *notational changes of non-summation indices must be uniformly applied to all terms in an equation*. These indices, on substitution, nominally give rise to four equations.

As a marker, *an index may be substituted by the elements of any set of four members* such as {1, 2, 3, 4}, {0, 1, 2, 3} or {x, y, z, t} and so on. So the equation above yields:

$$A'_1 = \frac{\partial x^\nu}{\partial x'^1} A_\nu \quad A'_2 = \frac{\partial x^\mu}{\partial x'^2} A_\mu \quad A'_3 = \frac{\partial x^\tau}{\partial x'^3} A_\tau \quad A'_4 = \frac{\partial x^\lambda}{\partial x'^4} A_\lambda$$

C2. <u>Tensors of the Second Rank</u>

C2.1. Contravariant Tensors

Consider: $A^{\mu\nu} = A^\mu B^\nu$; A^μ and B^ν are 4-vectors.

According to § C1.1 this satisfies the transformation:

$$A'^{\tau\sigma} = \frac{\partial x'^\sigma}{\partial x^\mu} A^\mu \frac{\partial x'^\tau}{\partial x^\nu} B^\nu = \frac{\partial x'^\sigma \partial x'^\tau}{\partial x^\mu \, \partial x^\nu} A^{\mu\nu}$$

$A^{\mu\nu}$ is a 16-vector (a tensor with sixteen components). Not all contravariant 16-vectors can be formed from two 4-vectors, however all satisfy this transformation.

C2.2. Covariant Tensors

Similarly, as above, a covariant tensor of the second rank may be formed by two covariant 4-vectors:

$A_{\mu\nu} = A_\mu B_\nu$; A_μ and B_ν are 4-vectors.

According to § C1.2 this tensor satisfies the transformation:

$$A'_{\sigma\tau} = \frac{\partial x^{\mu}}{\partial x'^{\sigma}} A_{\mu} \frac{\partial x^{\nu}}{\partial x'^{\tau}} B_{\nu} = \frac{\partial x^{\mu}}{\partial x'^{\sigma}} \frac{\partial x^{\nu}}{\partial x'^{\tau}} A_{\mu\nu}$$

Not all covariant 16-vectors can be formed from two 4-vectors, however they all satisfy this transformation.

C2.3 Mixed Tensors

Consider: $A_{\mu}^{\nu} = A_{\mu} B^{\nu}$:A_{μ} and B^{ν} are 4-vectors.

According to § C1.1 and § C1.2 this tensor satisfies the transformation:

$$A'^{\tau}_{\sigma} = \frac{\partial x^{\mu}}{\partial x'^{\sigma}} A_{\mu} \frac{\partial x'^{\tau}}{\partial x^{\nu}} B^{\nu} = \frac{\partial x^{\mu}}{\partial x'^{\sigma}} \frac{\partial x'^{\tau}}{\partial x^{\nu}} A_{\mu}^{\nu}$$

The tensor A_{μ}^{ν} is covariant with respect to μ and contravariant with respect to ν. As such it is said to be mixed.

C2.4 Symmetrical Tensors

Consider the equations:

$$A_{\mu\nu} = A_{\nu\mu}$$

$$A^{\mu\nu} = A^{\nu\mu}$$

and

$$A_{\mu}^{\nu} = A_{\nu}^{\mu}$$

The tensors in these equations are examples of symmetrical tensors of the second rank. For a tensor that is symmetrical with respect to two indices, interchange of these indices makes no difference to its value.

Of the sixteen quantities of a symmetrical tensor of the second rank, the four with repeated indices are of arbitrary quantities and of the remaining twelve; there are two identical sets of six quantities. Therefore ten independent quantities determine the symmetric tensor of the second rank so it is termed a 10-vector.

C2.5 Anti-Symmetrical Tensors

Anti-symmetrical tensors satisfy the following equations:

$$A_{\mu\nu} = -A_{\nu\mu}$$

$$A^{\mu\nu} = -A^{\nu\mu}$$

The indices that are interchanged must be either covariant or contravariant.

Of the sixteen quantities of an anti-symmetrical tensor of the second rank, the four with repeated indices are all zeroes. Of the remaining twelve, these may be divided into two sets each of six quantities with the elements of one set being of the same magnitude as, but of opposite signs to, elements of the other set. Thus there are six independent quantities to be determined: the anti-symmetrical tensor of the second rank is therefore termed a 6-vector.

C3. Basic Tensor Operations

C3.1 Outer Multiplication

In this operation, each component of a vector is multiplied by all components of another vector to form components of a new vector. The following are examples:

$$T_{\mu\nu\sigma} = A_{\mu\nu}B_{\sigma}$$

$$T_{\mu\nu\sigma\tau} = A_{\mu\nu}B_{\sigma\tau}$$

$$T_{\mu\nu}^{\sigma\tau} = A_{\mu\nu}B^{\sigma\tau}$$

This operation results in a tensor of rank equal to the sum of the ranks of the constitutive tensors. Proofs follow directly from the transformation equations of the constitutive tensors in § C1 and § C2. The construction of second rank tensors from first rank tensors in § C2 are examples of outer multiplication.

C3.2 Contraction of a Mixed Tensor

Consider the mixed tensor of the fourth rank $A_{\mu\nu}^{\sigma\tau}$.

For $\nu \equiv \tau = \alpha$ we have: $$A_{\mu\nu}^{\sigma\tau} = A_{\mu\alpha}^{\sigma\alpha}$$

That is, identify an index of the contravariant type with one of the covariant type. Next perform the summation over the index α, explicitly:

$$A_{\mu}^{\sigma} = \sum_{\alpha=1}^{4} A_{\mu\alpha}^{\sigma\alpha}$$

This results in a tensor of rank 2. Similarly:

$$A_{\mu}^{\sigma\tau} = A_{\mu}^{\mu\tau} = A^{\tau}$$

and $$A_{\mu}^{\tau} = A_{\mu}^{\mu} = A$$

Contraction implies a loss of information.

It is often convenient to treat scalars - constants or invariant functions of the coordinates - as tensors of zero rank.

C3.3 Inner Multiplication of Tensors of Different Types

Consider:

$$D_\nu = D^\sigma_{\mu\nu} = A_{\mu\nu}B^\sigma$$

Here, there is an outer multiplication followed by a contraction. This operation results in a tensor of a rank that is less than the sum of the ranks of the constitutive tensors.

C3.3.1 Note on the Inner Multiplication of Tensors

Consider, for example:

$$D^{\sigma\tau}_{\mu\nu} = A_{\mu\nu}B^{\sigma\tau}$$

Through a double contraction of the tensor $D^{\sigma\tau}_{\mu\nu}$ it becomes a scalar.

Conversely, if $A_{\mu\nu}B^{\mu\nu}$ is a scalar for any contravariant tensor $B^{\mu\nu}$, then $A_{\mu\nu}$ is a covariant tensor. The proof is as follows:

Consider:

$$A'_{\sigma\tau}\, B'^{\sigma\tau} = A_{\mu\nu}B^{\mu\nu}$$

Where $B'^{\sigma\tau}$ is a transform of $B^{\mu\nu}$, so:

$$B^{\mu\nu} = \frac{\partial x^\mu}{\partial x'^\sigma}\frac{\partial x^\nu}{\partial x'^\tau}B'^{\sigma\tau}$$

And:

$$A'_{\sigma\tau}B'^{\sigma\tau} = A_{\mu\nu}\frac{\partial x^\mu}{\partial x'^\sigma}\frac{\partial x^\nu}{\partial x'^\tau}B'^{\sigma\tau}$$

Which, for arbitrary $B'^{\sigma\tau}$ implies:

$$A'_{\sigma\tau} = \frac{\partial x^\mu}{\partial x'^\sigma}\frac{\partial x^\nu}{\partial x'^\tau}A_{\mu\nu}$$

That is, $A_{\mu\nu}$ is a covariant tensor, as was to be proved.

D. The Metric Tensor I

According to the considerations of § C1.1 and § C3.1, the product of the outer multiplication - $dx'^\tau dx'^\sigma$ - constitutes a contravariant tensor of the second rank. Therefore in light of the considerations of § C3.3.1, the metric given in equation (3) is a covariant tensor of the second rank as when it is multiplied by $dx'^\tau dx'^\sigma$ there results the scalar ds^2.

Being covariant, the metric tensor transforms as follows:

$$g'_{\sigma\tau} = \frac{\partial x^\mu}{\partial x'^\sigma}\frac{\partial x^\nu}{\partial x'^\tau} g_{\mu\nu}$$

Also, by interchanging indices in (3), it is clear that *the metric tensor is symmetric.*

The tensors of the metric may be considered as elements of a 4x4 matrix with the indices of the former corresponding to rows and columns of the latter.

If the cofactor of the metric - $g_{\alpha\sigma}$ - is divided by the latter's determinant - g, the *reciprocal metric* $g^{\sigma\tau}$ is formed. Because the covariant metric tensor is symmetric so is its cofactor and, therefore, so is its reciprocal.

Furthermore, the product of the metric and its reciprocal yields:

$$g_{\alpha\sigma}g^{\sigma\tau} = g_\alpha^\tau = 1 \qquad :\alpha = \tau$$
$$= 0 \qquad :\alpha \neq \tau$$

So g_α^τ is the *Kronecker delta,* δ_α^τ.

The quantity $g^{\mu\nu}$ is a contravariant tensor of the second rank - as multiplication with its reciprocal covariant tensor yields a scalar, so $g^{\mu\nu}$ transforms as follows:

$$g'^{\sigma\tau} = \frac{\partial x'^\sigma}{\partial x^\mu}\frac{\partial x'^\tau}{\partial x^\nu} g^{\mu\nu}$$

Formed by the product of two tensors, it follows that the Kronecker delta δ_α^τ is a mixed tensor of the second rank.

It also follows that summed over the four elements of the index μ or ν:

$$g_{\mu\nu}g^{\mu\nu} = \delta_\mu^\mu = \delta_\nu^\nu = 4$$

Now due to symmetry of $g'_{\sigma\tau}$ in σ and τ and of $g_{\mu\nu}$ in μ and ν:

$$g' = |g'_{\sigma\tau}| = \left|\frac{\partial x^\mu}{\partial x'^\sigma}\frac{\partial x^\nu}{\partial x'^\tau} g_{\mu\nu}\right| = \left|\frac{\partial x^\mu}{\partial x'^\sigma}\right|^2 g$$

Since the determinant of the metric in the special case of STR is −1, consider:

$$-g' = -g\left|\frac{\partial x^\mu}{\partial x'^\sigma}\right|^2$$

Thus we may, avoiding complex numbers, generally write:

$$\sqrt{-g'} = \sqrt{-g}\left|\frac{\partial x^\mu}{\partial x'^\sigma}\right|$$

Recalling the relationship of the determinant of the Jacobian matrix and its associated infinitesimal volumes:

$$d\tau' \stackrel{\text{def}}{=} \left|\frac{\partial x'^{\sigma}}{\partial x^{\mu}}\right| d\tau$$

where dτ is the differential 4-volume.

Multiplying the last two equations, we get:

$$\sqrt{-g'}\, d\tau' = \sqrt{-g}\, d\tau$$

Clearly g cannot vanish as then a finite volume may result from transforming an infinitesimal volume. Therefore, *g cannot change sign*. We shall *generally constrain* the choice of coordinate charts so that even in the absence of STR conditions:

$$\sqrt{-g} = \sqrt{-g'} = 1$$

which requires the Jacobian determinant to be unity:

$$\left|\frac{\partial x'^{\sigma}}{\partial x^{\mu}}\right| = 1$$

and yields: $$d\tau' = d\tau$$

which is quite acceptable as a transformation of an infinitesimal volume.

E. Formation of Tensors by the Metric

Multiplying the contravariant 4-vector transformation by the covariant metric:

$$g_{\mu\sigma} A^{\sigma} = g_{\mu\sigma} \frac{\partial x^{\sigma}}{\partial x^{\nu}} A^{\nu}$$

Changing indices μ to ν and τ to μ in equation (2) and substituting in the above:

$$g_{\mu\sigma} A^{\sigma} = \frac{\partial x^{\nu}}{\partial x^{\mu}} \frac{\partial x^{\nu}}{\partial x^{\sigma}} \frac{\partial x^{\sigma}}{\partial x^{\nu}} A^{\nu}$$

$$= \frac{\partial x^{\nu}}{\partial x^{\mu}} A^{\nu} \stackrel{\text{def}}{=} A_{\mu}$$

Since the summation index occurs in the numerator and $\partial x^{\nu}/\partial x^{\mu}$ is a factor of a covariant double tensor then the transformation is covariant and we designate the arbitrary quantity A^{ν} as A_{ν}, a covariant quantity of the same value.

Now consider: $$g_{\beta\sigma} A^{\tau\sigma} = g_{\beta\sigma} \frac{\partial x^{\sigma}}{\partial x^{\mu}} \frac{\partial x^{\tau}}{\partial x^{\nu}} A^{\mu\nu}$$

With (2) this yields:

$$g_{\beta\sigma}A^{\tau\sigma} = \frac{\partial x^{\mu}}{\partial x^{\beta}}\frac{\partial x^{\mu}}{\partial x^{\sigma}}\frac{\partial x^{\sigma}}{\partial x^{\mu}}\frac{\partial x^{\tau}}{\partial x^{\nu}}A^{\mu\nu} = \frac{\partial x^{\mu}}{\partial x^{\beta}}\frac{\partial x^{\tau}}{\partial x^{\nu}}A^{\nu}_{\mu} \stackrel{\text{def}}{=} A^{\tau}_{\beta}$$

as $\partial x^{\mu}/\partial x^{\beta}$ is a factor of a covariant double tensor and $\partial x^{\tau}/\partial x^{\nu}$ is a factor of a contravariant transformation. $A^{\mu\nu}$ is an arbitrary quantity that is now the object of a mixed transformation so it is correspondingly designated as A^{ν}_{μ}.

Another similar operation on the equation above yields:

$$A_{\alpha\beta} = g_{\alpha\tau}g_{\beta\sigma}A^{\tau\sigma} = g_{\alpha\tau}A^{\tau}_{\beta}$$

These operations are called 'lowering the index'.

Similarly, we may also 'raise the index' as follows:

$$A^{\nu} = g^{\nu\sigma}A_{\sigma}$$

$$A^{\nu}_{\alpha} = g^{\nu\beta}A_{\alpha\beta}$$

and

$$A^{\mu\nu} = g^{\mu\alpha}A^{\nu}_{\alpha} = g^{\mu\alpha}g^{\nu\beta}A_{\alpha\beta}$$

Also, it follows that: $A = g_{\mu\nu}A^{\mu\nu} = g^{\mu\nu}A_{\mu\nu}$

These last two operations are called metric contractions.

F. The Geodesic Equation

A trajectory that realizes, generally under constraints, the stationary magnitude of a line is called a geodesic. Under STR conditions, the trajectory of a freely moving body realizes the stationary magnitude of the linear element as a straight world-line; a constant velocity. Now we consider the general case.

Stationarity is determined by the use of variational methods:

$$\delta\int_{p}^{p'} ds = 0$$

Here δ implies an infinitesimal variation of the trajectory from the geodesic. The end-point events of the trajectory, P and P', are fixed.

Let the independent variable λ parameterize the family of 4-D surfaces intersecting the geodesic and other lines nearby that go through P and P'. The trajectories may then be considered as functions of λ and therefore the coordinates are also functions of λ.

Also, recalling (3), let: $$w^2 = \left(\frac{ds}{d\lambda}\right)^2 = g_{\mu\nu}\frac{dx^{\mu}}{d\lambda}\frac{dx^{\nu}}{d\lambda}$$

And: $$\delta(w^2) = 2w\delta w = \frac{\partial g_{\mu\nu}}{\partial x^\sigma}\frac{dx^\mu}{d\lambda}\frac{dx^\nu}{d\lambda}\delta x^\sigma + 2g_{\mu\nu}\frac{dx^\mu}{d\lambda}\frac{d(\delta x^\nu)}{d\lambda}$$

as $\delta(dx^\nu/d\lambda)=d(\delta x^\nu)/d\lambda$ [See Appendix § A]. Changing summation index ν to σ in the last term and rewriting:

$$\delta w = \frac{1}{2w}\frac{\partial g_{\mu\nu}}{\partial x^\sigma}\frac{dx^\mu}{d\lambda}\frac{dx^\nu}{d\lambda}\delta x^\sigma + \frac{g_{\mu\sigma}}{w}\frac{dx^\mu}{d\lambda}\frac{d(\delta x^\sigma)}{d\lambda}$$

With a change of variable and a partial integration of the last term above, we get:

$$\delta\int_{p}^{p'} ds = \delta\int_{\lambda_p}^{\lambda_{p'}} w d\lambda = \int_{\lambda_p}^{\lambda_{p'}} \delta w d\lambda = \int_{\lambda_p}^{\lambda_{p'}} \left[\frac{1}{2w}\frac{\partial g_{\mu\nu}}{\partial x^\sigma}\frac{dx^\mu}{d\lambda}\frac{dx^\nu}{d\lambda} - \frac{d}{d\lambda}\left(\frac{g_{\mu\sigma}}{w}\frac{dx^\mu}{d\lambda}\right)\right]\delta x^\sigma d\lambda$$

$$+ \left[\frac{g_{\mu\sigma}}{w}\frac{dx^\mu}{d\lambda}\delta x^\sigma\right]\Bigg|_{\lambda_p}^{\lambda_{p'}}$$

However, at P and P′, $\delta x^\sigma = 0$. Hence the last term of the equation above vanishes. Therefore, for arbitrary values of δx^σ, stationarity requires:

$$\frac{1}{2}\frac{\partial g_{\mu\nu}}{\partial x^\sigma}\frac{dx^\mu}{d\lambda}\frac{dx^\nu}{d\lambda} - \frac{d}{d\lambda}\left(g_{\mu\sigma}\frac{dx^\mu}{d\lambda}\right) = 0$$

With differentiation of the second term, this equation may be written as:

$$g_{\mu\sigma}\frac{d^2x^\mu}{d\lambda^2} + \frac{\partial g_{\mu\sigma}}{\partial x^\alpha}\frac{dx^\alpha}{d\lambda}\frac{dx^\mu}{d\lambda} - \frac{1}{2}\frac{\partial g_{\mu\nu}}{\partial x^\sigma}\frac{dx^\mu}{d\lambda}\frac{dx^\nu}{d\lambda} = 0$$

The notation of the summation indices is immaterial, so we may independently: change μ to α in the first term; change α to ν then interchange ν and μ in the second term; let μ and ν remain unchanged in the third term and write:

$$g_{\alpha\sigma}\frac{d^2x^\alpha}{d\lambda^2} + \frac{1}{2}\left[\frac{\partial g_{\mu\sigma}}{\partial x^\nu}\frac{dx^\nu}{d\lambda}\frac{dx^\mu}{d\lambda} + \frac{\partial g_{\nu\sigma}}{\partial x^\mu}\frac{dx^\mu}{d\lambda}\frac{dx^\nu}{d\lambda}\right] - \frac{1}{2}\frac{\partial g_{\mu\nu}}{\partial x^\sigma}\frac{dx^\mu}{d\lambda}\frac{dx^\nu}{d\lambda} = 0$$

Or: $$g_{\sigma\alpha}\frac{d^2x^\alpha}{d\lambda^2} + \frac{1}{2}\left[\frac{\partial g_{\mu\sigma}}{\partial x^\nu} + \frac{\partial g_{\nu\sigma}}{\partial x^\mu} - \frac{\partial g_{\mu\nu}}{\partial x^\sigma}\right]\frac{dx^\mu}{d\lambda}\frac{dx^\nu}{d\lambda} = 0$$

If s does not vanish on the trajectories, then let $\lambda = s$ and this equation may be rewritten as:

$$g_{\sigma\alpha}\frac{d^2x^\alpha}{ds^2} + [\mu\nu,\sigma]\frac{dx^\mu}{ds}\frac{dx^\nu}{ds} = 0 \qquad (4a)$$

where $$[\mu\nu,\sigma] \stackrel{\text{def}}{=} \frac{1}{2}\left(\frac{\partial g_{\mu\sigma}}{\partial x^{\nu}} + \frac{\partial g_{\nu\sigma}}{\partial x^{\mu}} - \frac{\partial g_{\mu\nu}}{\partial x^{\sigma}}\right) \tag{4b}$$

On the right of equation (4b), the sum of the first two terms in the brackets is symmetric in μ and ν and so is the third term, therefore so is $[\mu\nu, \sigma]$.

Multiply equation (4a) by the contravariant metric tensor $g^{\sigma\tau}$:

$$g^{\sigma\tau} g_{\sigma\alpha} \frac{d^2x^{\alpha}}{ds^2} + g^{\sigma\tau}[\mu\nu,\sigma]\frac{dx^{\mu}}{ds}\frac{dx^{\nu}}{ds} = 0$$

Recall: $$g^{\sigma\tau} g_{\sigma\alpha} = \delta^{\tau}_{\alpha} = 1 \quad : \text{for } \alpha = \tau$$
$$= 0 \quad : \text{for } \alpha \neq \tau$$

Therefore: $$\frac{d^2x^{\tau}}{ds^2} + g^{\sigma\tau}[\mu\nu,\sigma]\frac{dx^{\mu}}{ds}\frac{dx^{\nu}}{ds} = 0 \tag{4c}$$

and $$\frac{d^2x^{\tau}}{ds^2} + \{\mu\nu,\tau\}\frac{dx^{\mu}}{ds}\frac{dx^{\nu}}{ds} = 0 \tag{5}$$

where $$\{\mu\nu, \tau\} \stackrel{\text{def}}{=} g^{\sigma\tau}[\mu\nu,\sigma] \tag{6}$$

Furthermore, letting:

$$\Gamma^{\tau}_{\mu\nu} \stackrel{\text{def}}{=} -\{\mu\nu,\tau\} = -g^{\sigma\tau}[\mu\nu,\sigma] = -\frac{g^{\sigma\tau}}{2}\left(\frac{\partial g_{\mu\sigma}}{\partial x^{\nu}} + \frac{\partial g_{\nu\sigma}}{\partial x^{\mu}} - \frac{\partial g_{\mu\nu}}{\partial x^{\sigma}}\right) \tag{7 *}$$

we may write: $$\frac{d^2x^{\tau}}{ds^2} = \Gamma^{\tau}_{\mu\nu}\frac{dx^{\mu}}{ds}\frac{dx^{\nu}}{ds} \qquad \tau = 1,2,3,4 \tag{8}$$

Equations (4c), (5) and (8) are various forms of the *geodesic equation*.

Equation (8) implies that, for $\Gamma^{\tau}_{\mu\nu} = 0$, the motion is a constant velocity: there is no gravitational acceleration. This implies that the $\Gamma^{\tau}_{\mu\nu}$ are *components of the gravitational field*. As $[\mu\nu, \sigma]$ is symmetrical in μ and ν, so are $\{\mu\nu, \tau\}$ and $\Gamma^{\tau}_{\mu\nu}$.

So the $g_{\mu\nu}$, along with determining the metrical properties of a chart applied to a region, also describe the region's gravitational field *with reference to the applied chart*. This feature distinguishes the gravitational field from other fields.[17]

It follows that for an observer fixed in her chart with both 'riding' on a geodesic: the $\partial g_{\mu\nu}/\partial x^{\sigma}$, the $\Gamma^{\tau}_{\mu\nu}$, the 4-D acceleration - d^2x^{τ}/ds^2 - and its embedded 3-D acceleration - Newton's field intensity - all vanish *relative to her* (in a free fall) *and her chart*[18]. Her chart becomes Minkowskian with constant $g_{\mu\nu}$. [cf. § A ¶7].

* Equation (7) is Einstein's definition of $\Gamma^{\tau}_{\mu\nu}$. This is the negative of the connection coefficient later used by Misner, Thorne and Wheeler in their influential work *Gravitation* that consolidated the foundation of the 'geometrization' of physics - an ultimately sterile process that led to the concept of the 'spacetime fabric'.

G. Tensor Calculus I

Here we consider tensors formed by differentiating tensors.

G1. Gradient 4-Vectors

Given $\varphi(x^\mu)$, an invariant function of the coordinates, then:

$$\psi = \frac{d\varphi}{ds} = \frac{\partial\varphi}{\partial x^\mu}\frac{dx^\mu}{ds}$$

being a quotient of invariants, is also invariant. However,

$$A_\mu = \frac{\partial\varphi}{\partial x^\mu}$$

is a covariant 4-vector as dx^μ/ds is a contravariant 4-vector and ψ is an invariant [cf. C3.3.1]. A_μ is the 4-D gradient of $\varphi(x^\mu)$.

Let:

$$X = \frac{d\psi}{ds}$$
$$= \frac{d}{ds}\left(A_\mu \frac{dx^\mu}{ds}\right)$$

X is also an invariant. Differentiating:

$$X = \frac{\partial A_\mu}{\partial x^\nu}\frac{dx^\nu}{ds}\frac{dx^\mu}{ds} + A_\mu\frac{d^2x^\mu}{ds^2}$$

Changing the summation index of the second term μ to τ and simultaneously solving with the geodesic equation (5) yield:

$$X = \left[\frac{\partial A_\mu}{\partial x^\nu} - \{\mu\nu, \tau\}A_\tau\right]\frac{dx^\nu}{ds}\frac{dx^\mu}{ds}$$
$$= A_{\mu\nu}\frac{dx^\nu}{ds}\frac{dx^\mu}{ds}$$

where:

$$A_{\mu\nu} = \frac{\partial A_\mu}{\partial x^\nu} - \{\mu\nu,\tau\}A_\tau \qquad (9)$$

$A_{\mu\nu}$ is a covariant 16-vector as X is invariant and the product $(dx^\mu/ds)(dx^\nu/ds)$ is a contravariant 16-vector. $A_{\mu\nu}$ is the ordinary derivative of the covariant 4-vector A_μ 'corrected' by $\{\mu\nu, \tau\}A_\tau$ to ensure geodicity. $A_{\mu\nu}$ is the *extension* of A_μ along a geodesic. $A_{\mu\nu}$ is also called the *covariant derivative* of the covariant 4-vector A_μ.

G2. A Certain Type of Non-Gradient Covariant 4-Vectors

In developing the extension of the covariant 4-vector, the A_μ previously considered are components of a gradient 4-vector. Now multiply equation (9) by the invariant Ψ:

$$\psi A_{\mu\nu} = \psi\frac{\partial A_\mu}{\partial x^\nu} - \{\mu\nu,\tau\}A_\tau\psi = \psi\frac{\partial^2\varphi}{\partial x^\nu\partial x^\mu} - \{\mu\nu,\tau\}\frac{\partial\varphi}{\partial x^\tau}\psi$$

We may also write:

$$\psi A_{\mu\nu} + \frac{\partial\psi}{\partial x^\nu}\frac{\partial\varphi}{\partial x^\mu} = \psi\frac{\partial^2\varphi}{\partial x^\nu\partial x^\mu} + \frac{\partial\psi}{\partial x^\nu}\frac{\partial\varphi}{\partial x^\mu} - \{\mu\nu,\tau\}\frac{\partial\varphi}{\partial x^\tau}\psi$$

Rewrite this equation as:

$$\psi A_{\mu\nu} + \frac{\partial\psi}{\partial x^\nu}\frac{\partial\varphi}{\partial x^\mu} = \frac{\partial}{\partial x^\nu}\left(\psi\frac{\partial\varphi}{\partial x^\mu}\right) - \{\mu\nu,\tau\}\frac{\partial\varphi}{\partial x^\tau}\psi$$

As both terms on the left hand side are covariant tensors of the second rank, so is the expression on the right. Hence, the expression on the right may be rewritten as the covariant extension:

$$B_{\mu\nu} = \frac{\partial B_\mu}{\partial x^\nu} - \{\mu\nu,\tau\}B_\tau$$

Here

$$B_\mu = \psi\frac{\partial\varphi}{\partial x^\mu}$$

is not a gradient 4-vector.

H. The Metric Tensor II

Recall the rule for the differentiation of determinants: for a matrix M with determinant m;

$$\frac{dm}{dM} \overset{\text{def}}{=} (mM^{-1})^T$$

where T denotes the transpose.

Applying this rule to $g_{\mu\nu}$ yields:

$$dg = g^{\mu\nu}g\,dg_{\mu\nu} \qquad (10a)$$

Recall:

$$g_{\mu\nu}g^{\mu\nu} = 4$$

This leads to:

$$g_{\mu\nu}dg^{\mu\nu} + g^{\mu\nu}dg_{\mu\nu} = 0$$

So we may also write:

$$dg = -g_{\mu\nu}g\,dg^{\mu\nu} \qquad (10b)$$

Equations (10a) and (10b) yield:

$$d(-g)/(-g) = g^{\mu\nu}dg_{\mu\nu} = -g_{\mu\nu}dg^{\mu\nu}$$

and: $$d\{\ln(-g)\} = g^{\mu\nu}dg_{\mu\nu} = -\, g_{\mu\nu}dg^{\mu\nu}$$

We may also write:

$$\frac{\partial\{\ln(-g)\}}{\partial x^\sigma} = g^{\mu\nu}\frac{\partial g_{\mu\nu}}{\partial x^\sigma} = -\, g_{\mu\nu}\frac{\partial g^{\mu\nu}}{\partial x^\sigma} \quad (11a)$$

Recall that:

$$g_{\mu\sigma}g^{\nu\sigma} = \delta^\nu_\mu = 1 \quad :\mu = \nu$$
$$= 0 \quad :\mu \neq \nu$$

That implies:

$$g_{\mu\sigma}\frac{\partial g^{\nu\sigma}}{\partial x^\lambda} = -\, g^{\nu\sigma}\frac{\partial g_{\mu\sigma}}{\partial x^\lambda} \quad (11b)$$

and:

$$g^{\tau\mu}g_{\mu\sigma}\frac{\partial g^{\nu\sigma}}{\partial x^\lambda} = \delta^\tau_\sigma\frac{\partial g^{\nu\sigma}}{\partial x^\lambda} = -\, g^{\tau\mu}g^{\nu\sigma}\frac{\partial g_{\mu\sigma}}{\partial x^\lambda}$$

In the last term, change the notation of the summation indices μ and σ to α and β, respectively. Contracting the middle term, let $\sigma = \tau$. So we get:

$$\frac{\partial g^{\nu\tau}}{\partial x^\lambda} = -\, g^{\tau\alpha}g^{\nu\beta}\frac{\partial g_{\alpha\beta}}{\partial x^\lambda} \quad (12)$$

Now change the non-summation index τ to μ:

$$\frac{\partial g^{\mu\nu}}{\partial x^\lambda} = -\, g^{\mu\alpha}g^{\nu\beta}\frac{\partial g_{\alpha\beta}}{\partial x^\lambda} \quad (13)$$

Recalling (4b), we may write:

$$[\alpha\sigma,\beta] = \frac{1}{2}\left(\frac{\partial g_{\alpha\beta}}{\partial x^\sigma} + \frac{\partial g_{\sigma\beta}}{\partial x^\alpha} - \frac{\partial g_{\alpha\sigma}}{\partial x^\beta}\right)$$

and: $$[\beta\sigma,\alpha] = \frac{1}{2}\left(\frac{\partial g_{\beta\alpha}}{\partial x^\sigma} + \frac{\partial g_{\sigma\alpha}}{\partial x^\beta} - \frac{\partial g_{\beta\sigma}}{\partial x^\alpha}\right)$$

Symmetry of $g_{\mu\nu}$ in μ and ν implies that, in the round brackets of both equations, the first terms are symmetrical in α and β and the sums of the last two terms are anti-symmetrical in the same indices and cancel in the sum of these equations, so:

$$[\alpha\sigma,\beta] + [\beta\sigma,\alpha] = \frac{\partial g_{\alpha\beta}}{\partial x^\sigma} \quad (14)$$

With equation (13) this becomes:

$$-g^{\mu\alpha}g^{\nu\beta}\frac{\partial g_{\alpha\beta}}{\partial x^\sigma} = \frac{\partial g^{\mu\nu}}{\partial x^\sigma} = -\, g^{\mu\alpha}g^{\nu\beta}\{[\alpha\sigma,\beta] + [\beta\sigma,\alpha]\}$$

Recalling (6) yields: $$\frac{\partial g^{\mu\nu}}{\partial x^\sigma} = -\, g^{\mu\alpha}\{\alpha\sigma,\nu\} - g^{\nu\beta}\{\beta\sigma,\mu\}$$

With notational changes of both summation indices α and β to τ, we get:

$$\frac{\partial g^{\mu\nu}}{\partial x^{\sigma}} = -g^{\mu\tau}\{\tau\sigma,\nu\} - g^{\nu\tau}\{\tau\sigma,\mu\} \tag{15}$$

Applying (7) yields:

$$\frac{\partial g^{\mu\nu}}{\partial x^{\sigma}} = g^{\mu\tau}\Gamma^{\nu}_{\tau\sigma} + g^{\nu\tau}\Gamma^{\mu}_{\tau\sigma} \tag{16}$$

Rewriting (11a):

$$\frac{1}{2}\frac{\partial}{\partial x^{\sigma}}\{\ln(-g)\} = \frac{\partial}{\partial x^{\sigma}}(\ln\sqrt{-g}) = \frac{1}{2}\, g^{\mu\nu}\frac{\partial g_{\mu\nu}}{\partial x^{\sigma}} = -\frac{1}{2}\, g_{\mu\nu}\frac{\partial g^{\mu\nu}}{\partial x^{\sigma}}$$

Differentiating the second expression and substituting from (15):

$$\frac{1}{\sqrt{-g}}\frac{\partial\sqrt{-g}}{\partial x^{\sigma}} = \frac{1}{2}\, g^{\mu\nu}\frac{\partial g_{\mu\nu}}{\partial x^{\sigma}} = -\frac{1}{2}\, g_{\mu\nu}\frac{\partial g^{\mu\nu}}{\partial x^{\sigma}} = \frac{1}{2} g_{\mu\nu}[g^{\mu\tau}\{\tau\sigma,\nu\} + g^{\nu\tau}\{\tau\sigma,\mu\}]$$

Expanding the expression on the right and recalling that $g^{\sigma\tau}g_{\sigma\alpha} = \delta^{\tau}_{\alpha}$:

$$\frac{1}{\sqrt{-g}}\frac{\partial\sqrt{-g}}{\partial x^{\sigma}} = \frac{1}{2}\, g^{\mu\nu}\frac{\partial g_{\mu\nu}}{\partial x^{\sigma}} = -\frac{1}{2}\, g_{\mu\nu}\frac{\partial g^{\mu\nu}}{\partial x^{\sigma}} = \frac{1}{2}[\delta^{\tau}_{\nu}\{\tau\sigma,\nu\} + \delta^{\tau}_{\mu}\{\tau\sigma,\mu\}]$$

$$= \frac{1}{2}[\{\nu\sigma,\nu\} + \{\mu\sigma,\mu\}]$$

after contraction. On the right, change of summation index of ν to μ yields:

$$\{\mu\sigma,\mu\} = \frac{1}{\sqrt{-g}}\frac{\partial\sqrt{-g}}{\partial x^{\sigma}} = \frac{1}{2}\, g^{\mu\nu}\frac{\partial g_{\mu\nu}}{\partial x^{\sigma}} = -\frac{1}{2}\, g_{\mu\nu}\frac{\partial g^{\mu\nu}}{\partial x^{\sigma}} = \frac{\partial}{\partial x^{\sigma}}(\ln\sqrt{-g}) \tag{17}$$

We shall have much use of these relationships further on.

I. Tensor Calculus II

I1. Curl of a Covariant 4-Vector

Consider equation (9):

$$A_{\mu\nu} = \frac{\partial A_{\mu}}{\partial x^{\nu}} - \{\mu\nu,\tau\}A_{\tau}$$

Symmetry of $\{\mu\nu,\tau\}$ in μ and ν implies:

$$B_{\mu\nu} = A_{\mu\nu} - A_{\nu\mu} = \frac{\partial A_{\mu}}{\partial x^{\nu}} - \frac{\partial A_{\nu}}{\partial x^{\mu}} \tag{18}$$

$B_{\mu\nu}$ is the curl of A_{μ} and is anti-symmetrical as interchange of indices shows.

The curl is a measure of the rotational intensity of a tensor.

I2. Extension of Tensors of the Second Rank

I2.1 Covariant Tensors

Consider the sum of the outer products:

$$A_{\mu\nu\sigma} = A_{\mu\sigma}B_\nu + B_{\nu\sigma}A_\mu$$

where $A_{\mu\sigma}$ and $B_{\nu\sigma}$ are extensions of the covariant 4-vectors A_μ and B_ν, respectively. Substituting equation (9) yields:

$$A_{\mu\nu\sigma} = \left(\frac{\partial A_\mu}{\partial x^\sigma} - \{\sigma\mu,\tau\}A_\tau\right)B_\nu + \left(\frac{\partial B_\nu}{\partial x^\sigma} - \{\sigma\nu,\tau\}B_\tau\right)A_\mu$$

Expanding: $$A_{\mu\nu\sigma} = \frac{\partial A_\mu}{\partial x^\sigma}B_\nu + A_\mu\frac{\partial B_\nu}{\partial x^\sigma} - \{\sigma\mu,\tau\}A_\tau B_\nu - \{\sigma\nu,\tau\}A_\mu B_\tau$$

Letting $A_{\mu\nu} = A_\mu B_\nu$ then $A_{\mu\nu\sigma}$ may be expressed as:

$$A_{\mu\nu\sigma} = \frac{\partial A_{\mu\nu}}{\partial x^\sigma} - \{\sigma\mu,\tau\}A_{\tau\nu} - \{\sigma\nu,\tau\}A_{\mu\tau} \qquad (19)$$

$A_{\mu\nu\sigma}$ is the extension of $A_{\mu\nu}$ a covariant tensor of the second rank. $A_{\mu\nu\sigma}$ is linear in $A_{\mu\nu}$ and homogenous. Since $A_{\mu\nu}$ is a covariant tensor, so is $A_{\mu\nu\sigma}$. It is a tensor of the third rank.

I2.1.1 Anti-Symmetrical Covariant Tensors

In equation (19), let $A_{\mu\nu}$ be an anti-symmetric tensor of the second rank, a 6-vector. Consider the following cyclic rotation of indices:

$$A_{\mu\nu\sigma} = \frac{\partial A_{\mu\nu}}{\partial x^\sigma} - \{\sigma\mu,\tau\}A_{\tau\nu} - \{\sigma\nu,\tau\}A_{\mu\tau}$$

$$A_{\nu\sigma\mu} = \frac{\partial A_{\nu\sigma}}{\partial x^\mu} - \{\mu\nu,\tau\}A_{\tau\sigma} - \{\mu\sigma,\tau\}A_{\nu\tau}$$

$$A_{\sigma\mu\nu} = \frac{\partial A_{\sigma\mu}}{\partial x^\nu} - \{\nu\sigma,\tau\}A_{\tau\mu} - \{\nu\mu,\tau\}A_{\sigma\tau}$$

Since $\{\mu\nu,\tau\}$ is symmetric in μ and ν and $A_{\tau\sigma}$ is anti-symmetric in τ and σ, then in the summation of all the equations there results, on the right, mutual cancellation of all the second and third terms of the three equations above. Hence:

$$B_{\mu\nu\sigma} = A_{\mu\nu\sigma} + A_{\nu\sigma\mu} + A_{\sigma\mu\nu}$$

$$= \frac{\partial A_{\mu\nu}}{\partial x^\sigma} + \frac{\partial A_{\nu\sigma}}{\partial x^\mu} + \frac{\partial A_{\sigma\mu}}{\partial x^\nu} \qquad (20)$$

$B_{\mu\nu\sigma}$ is the extension of $A_{\mu\nu}$ and is anti-symmetric because so is $A_{\mu\nu}$.

I2.2 Contravariant Tensors

Multiplying (19) by $g^{\mu\alpha}g^{\nu\beta}$:

$$g^{\mu\alpha}g^{\nu\beta}A_{\mu\nu\sigma} = g^{\mu\alpha}g^{\nu\beta}\left[\frac{\partial A_{\mu\nu}}{\partial x^{\sigma}} - \{\sigma\mu,\tau\}A_{\tau\nu} - \{\sigma\nu,\tau\}A_{\mu\tau}\right] \tag{19a}$$

Rewrite the first term of the expanded right hand side of this equation as:

$$g^{\mu\alpha}g^{\nu\beta}\frac{\partial A_{\mu\nu}}{\partial x^{\sigma}} = \frac{\partial}{\partial x^{\sigma}}\left(g^{\mu\alpha}g^{\nu\beta}A_{\mu\nu}\right) - g^{\mu\alpha}\frac{\partial g^{\nu\beta}}{\partial x^{\sigma}}A_{\mu\nu} - g^{\nu\beta}\frac{\partial g^{\mu\alpha}}{\partial x^{\sigma}}A_{\mu\nu} \tag{19b}$$

Raise the indices of the first term on the right of the equation above and apply (15) to its last two terms. Then raise the indices on the left of the expanded equation (19a) and substitute the modified equation (19b) into it:

$$\begin{aligned} A^{\alpha\beta}_{\sigma} = {} & \frac{\partial A^{\alpha\beta}}{\partial x^{\sigma}} + g^{\mu\alpha}[g^{\nu\tau}\{\tau\sigma,\beta\} + g^{\beta\tau}\{\tau\sigma,\nu\}]A_{\mu\nu} \\ & + g^{\nu\beta}[g^{\mu\tau}\{\tau\sigma,\alpha\} + g^{\alpha\tau}\{\tau\sigma,\mu\}]A_{\mu\nu} \\ & - g^{\mu\alpha}g^{\nu\beta}\{\sigma\mu,\tau\}A_{\tau\nu} - g^{\mu\alpha}g^{\nu\beta}\{\sigma\nu,\tau\}A_{\mu\tau} \end{aligned}$$

Raise indices in the expanded second and third terms:

$$\begin{aligned} A^{\alpha\beta}_{\sigma} = {} & \frac{\partial A^{\alpha\beta}}{\partial x^{\sigma}} + \{\tau\sigma,\beta\}A^{\alpha\tau} + g^{\mu\alpha}g^{\beta\tau}\{\tau\sigma,\nu\}A_{\mu\nu} + \{\tau\sigma,\alpha\}A^{\tau\beta} \\ & + g^{\nu\beta}g^{\alpha\tau}\{\tau\sigma,\mu\}A_{\mu\nu} - g^{\mu\alpha}g^{\nu\beta}\{\sigma\mu,\tau\}A_{\tau\nu} - g^{\mu\alpha}g^{\nu\beta}\{\sigma\nu,\tau\}A_{\mu\tau} \end{aligned}$$

With notational interchange of summation indices τ and μ and because of the symmetry of $g^{\mu\nu}$ and $\{\mu\nu, .\}$ in μ and ν, then the sixth term on the right becomes:

$$-g^{\alpha\tau}g^{\nu\beta}\{\tau\sigma,\mu\}A_{\mu\nu}$$

Likewise with interchange of ν and τ, the seventh term becomes:

$$-g^{\mu\alpha}g^{\beta\tau}\{\tau\sigma,\nu\}A_{\mu\nu}$$

Hence, the third and the seventh terms cancel each other. So do the fifth and the sixth terms. And the extension of the contravariant tensor of the second rank is:

$$A^{\alpha\beta}_{\sigma} = \frac{\partial A^{\alpha\beta}}{\partial x^{\sigma}} + \{\sigma\tau,\beta\}A^{\alpha\tau} + \{\sigma\tau,\alpha\}A^{\tau\beta} \tag{21}$$

I2.3 Mixed Tensors

Interchanging σ and ν then ν and μ in equation (19) and multiplying by $g^{\nu\alpha}$:

$$g^{\nu\alpha}A_{\nu\sigma\mu} = g^{\nu\alpha}\frac{\partial A_{\nu\sigma}}{\partial x^{\mu}} - g^{\nu\alpha}\{\mu\nu,\tau\}A_{\tau\sigma} - g^{\nu\alpha}\{\mu\sigma,\tau\}A_{\nu\tau}$$

Rewriting the first term on the right:

$$g^{\nu\alpha}A_{\nu\sigma\mu} = \frac{\partial}{\partial x^{\mu}}(g^{\nu\alpha}A_{\nu\sigma}) - \frac{\partial g^{\nu\alpha}}{\partial x^{\mu}}A_{\nu\sigma} - g^{\nu\alpha}\{\mu\nu,\tau\}A_{\tau\sigma} - g^{\nu\alpha}\{\mu\sigma,\tau\}A_{\nu\tau}$$

Raising the indices on the left and in the first term on the right and recalling equation (15):

$$A^{\alpha}_{\sigma\mu} = \frac{\partial A^{\alpha}_{\sigma}}{\partial x^{\mu}} + g^{\nu\tau}\{\tau\mu,\alpha\}A_{\nu\sigma} + g^{\alpha\tau}\{\tau\mu,\nu\}A_{\nu\sigma} - g^{\nu\alpha}\{\mu\nu,\tau\}A_{\tau\sigma} - g^{\nu\alpha}\{\mu\sigma,\tau\}A_{\nu\tau}$$

Interchange σ and μ and then raise the indices in the second and last terms:

$$A^{\alpha}_{\mu\sigma} = \frac{\partial A^{\alpha}_{\mu}}{\partial x^{\sigma}} + \{\tau\sigma,\alpha\}A^{\tau}_{\mu} + g^{\alpha\tau}\{\tau\sigma,\nu\}A_{\nu\mu} - g^{\nu\alpha}\{\sigma\nu,\tau\}A_{\tau\mu} - \{\sigma\mu,\tau\}A^{\alpha}_{\tau}$$

Interchange of summation indices τ and ν in the third term reveals its cancellation with the fourth term and we arrive at the extension of the mixed tensor of the second rank:

$$A^{\alpha}_{\mu\sigma} = \frac{\partial A^{\alpha}_{\mu}}{\partial x^{\sigma}} - \{\sigma\mu,\tau\}A^{\alpha}_{\tau} + \{\sigma\tau,\alpha\}A^{\tau}_{\mu} \qquad (22)$$

I3. Divergence of Tensors

Divergence is the volume density of the outward flux of a tensor field through an infinitesimal 4-volume around a point-event. A positive divergence indicates a source, a negative divergence implies a sink, and a vanishing divergence means that there are no sources or sinks. The divergence is determined by the metric contraction of the extension of the tensor.

I3.1 Contravariant 4-Vector

Multiplying equation (9) by $g^{\mu\nu}$, rewriting the first term and recalling (7) yield:

$$g^{\mu\nu}A_{\mu\nu} = \frac{\partial}{\partial x^{\nu}}(g^{\mu\nu}A_{\mu}) - A_{\mu}\frac{\partial g^{\mu\nu}}{\partial x^{\nu}} - \frac{1}{2}g^{\tau\alpha}g^{\mu\nu}\left(\frac{\partial g_{\mu\alpha}}{\partial x^{\nu}} + \frac{\partial g_{\nu\alpha}}{\partial x^{\mu}} - \frac{\partial g_{\mu\nu}}{\partial x^{\alpha}}\right)A_{\tau} \qquad (23a)$$

In the expanded last term, interchange of summation indices μ and ν of its second term reveals its equality with the first, so we write the last term above as:

$$-g^{\tau\alpha}g^{\mu\nu}\frac{\partial g_{\mu\alpha}}{\partial x^{\nu}}A_{\tau}+\frac{1}{2}g^{\tau\alpha}g^{\mu\nu}\frac{\partial g_{\mu\nu}}{\partial x^{\alpha}}A_{\tau}$$

In the above expression, applying (13) to the first term and substituting from equation (17) in the second term yield:

$$\frac{\partial g^{\tau\nu}}{\partial x^{\nu}}A_{\tau}+\frac{g^{\tau\alpha}A_{\tau}\partial\sqrt{-g}}{\sqrt{-g}\ \partial x^{\alpha}}$$

Changing summation indices τ to μ in the first term and α to ν in the last term and then replacing the last term of equation (23a) by the resulting expression yield:

$$g^{\mu\nu}A_{\mu\nu}=\frac{\partial}{\partial x^{\nu}}(g^{\mu\nu}A_{\mu})+\frac{g^{\tau\nu}A_{\tau}\partial\sqrt{-g}}{\sqrt{-g}\ \partial x^{\nu}}$$

Now metrically contracting the term on the left and raising indices on the right yield:

$$A=\frac{\partial A^{\nu}}{\partial x^{\nu}}+\frac{A^{\nu}}{\sqrt{-g}}\frac{\partial\sqrt{-g}}{\partial x^{\nu}}$$

$$=\frac{1}{\sqrt{-g}}\left(\sqrt{-g}\frac{\partial A^{\nu}}{\partial x^{\nu}}+A^{\nu}\frac{\partial\sqrt{-g}}{\partial x^{\nu}}\right)$$

$$=\frac{1}{\sqrt{-g}}\frac{\partial}{\partial x^{\nu}}(A^{\nu}\sqrt{-g}) \qquad (23b)$$

This is the divergence of the contravariant 4-vector.

With constant $g_{\mu\nu}$, equation (23a) yields the divergence of a vector field. It generalizes the concept of divergence to tensors. This parallels the reduction - in STR conditions - of the covariant derivative to the ordinary derivative.

13.2 Contravariant Tensors of the Second Rank

Contracting equation (21) with respect to β and σ, let σ = β. This yields the divergence of the contravariant tensor of the second rank:

$$A^{\alpha}=\frac{\partial A^{\alpha\beta}}{\partial x^{\beta}}+\{\beta\tau,\beta\}A^{\alpha\tau}+\{\beta\tau,\alpha\}A^{\tau\beta}$$

Also, recalling (17) yields the alternative form:

$$A^{\alpha}=\frac{\partial A^{\alpha\beta}}{\partial x^{\beta}}+\frac{A^{\alpha\tau}}{\sqrt{-g}}\frac{\partial\sqrt{-g}}{\partial x^{\tau}}+\{\beta\tau,\alpha\}A^{\tau\beta} \qquad (24)$$

13.2.1 Anti-symmetrical Contravariant Tensors of the Second Rank

In equation (24), let $A^{\alpha\beta}$ be an anti-symmetrical tensor. Because of the symmetry of $\{\beta\tau,\alpha\}$ in the summation indices β and τ and the anti-symmetry of $A^{\tau\beta}$ in the same indices, then the third term vanishes. This occurs, as in the expanded third term, terms that have different elements for β and τ have their negatives and those terms with repeated elements vanish. Now changing τ to β:

$$A^{\alpha} = \frac{\partial A^{\alpha\beta}}{\partial x^{\beta}} + \frac{A^{\alpha\beta}}{\sqrt{-g}}\frac{\partial\sqrt{-g}}{\partial x^{\beta}}$$

We may rewrite this as:

$$A^{\alpha} = \frac{1}{\sqrt{-g}}\left[\sqrt{-g}\frac{\partial A^{\alpha\beta}}{\partial x^{\beta}} + A^{\alpha\beta}\frac{\partial\sqrt{-g}}{\partial x^{\beta}}\right]$$

$$= \frac{1}{\sqrt{-g}}\frac{\partial}{\partial x^{\beta}}\left(\sqrt{-g}A^{\alpha\beta}\right) \qquad (25)$$

This expresses the divergence of the contravariant 6-vector.

13.3 Mixed Tensors of the Second Rank

Contracting equation (22) with respect to α and σ, let $\alpha = \sigma$:

$$A_{\mu} = A^{\sigma}_{\sigma\mu} = \frac{\partial A^{\sigma}_{\mu}}{\partial x^{\sigma}} - \{\sigma\mu,\tau\}A^{\sigma}_{\tau} + \{\sigma\tau,\sigma\}A^{\tau}_{\mu}$$

Substituting from (17) in the last term and multiplying the equation by $\sqrt{-g}$:

$$\sqrt{-g}A_{\mu} = \sqrt{-g}\frac{\partial A^{\sigma}_{\mu}}{\partial x^{\sigma}} - \sqrt{-g}\{\sigma\mu,\tau\}A^{\sigma}_{\tau} + \frac{\partial\sqrt{-g}}{\partial x^{\tau}}A^{\tau}_{\mu}$$

In the last term, change summation indices τ to σ and rewrite the equation:

$$\sqrt{-g}A_{\mu} = \frac{\partial}{\partial x^{\sigma}}\left(\sqrt{-g}A^{\sigma}_{\mu}\right) - \{\sigma\mu,\tau\}\sqrt{-g}A^{\sigma}_{\tau} \qquad (26)$$

This equation gives the divergence A_{μ} of the mixed tensor of the second rank.

Recalling equation (6) we may introduce the symmetrical contravariant tensor $A^{\rho\sigma} = g^{\rho\tau}A^{\sigma}_{\tau}$ in the last term of equation (26) as follows:

$$\{\sigma\mu,\tau\}\sqrt{-g}A^{\sigma}_{\tau} = g^{\rho\tau}[\sigma\mu,\rho]\sqrt{-g}A^{\sigma}_{\tau}$$

$$= [\sigma\mu,\rho]\sqrt{-g}A^{\rho\sigma}$$

Substituting from equation (4b) yields:

$$\{\sigma\mu,\tau\}\sqrt{-g}A^{\sigma}_{\tau} = \left(\frac{\partial g_{\sigma\rho}}{\partial x^{\mu}} + \frac{\partial g_{\mu\rho}}{\partial x^{\sigma}} - \frac{\partial g_{\sigma\mu}}{\partial x^{\rho}}\right)\frac{\sqrt{-g}A^{\rho\sigma}}{2}$$

In light of the discussion in the first paragraph of § I.3.2.1, given the anti-symmetry of the difference of the last two terms in the round brackets in σ and ρ, then their product with the symmetrical $A^{\rho\sigma}$ vanishes. Hence:

$$\{\sigma\mu,\tau\}\sqrt{-g}A^{\sigma}_{\tau} = \frac{\partial g_{\sigma\rho}}{\partial x^{\mu}}\frac{\sqrt{-g}A^{\rho\sigma}}{2}$$

Substituting in (26):

$$\sqrt{-g}A_{\mu} = \frac{\partial}{\partial x^{\sigma}}\left(\sqrt{-g}A^{\sigma}_{\mu}\right) - \frac{\partial g_{\rho\sigma}}{\partial x^{\mu}}\frac{\sqrt{-g}A^{\rho\sigma}}{2} \tag{27}$$

This equation also yields the divergence A_{μ} of the mixed tensor.

Introduce the symmetrical covariant tensor $A_{\alpha\beta}$ by means of substituting $A^{\rho\sigma}$ with $g^{\rho\alpha}g^{\sigma\beta}A_{\alpha\beta}$ in equation (27):

$$\sqrt{-g}A_{\mu} = \frac{\partial}{\partial x^{\sigma}}\left(\sqrt{-g}A^{\sigma}_{\mu}\right) - \frac{\partial g_{\rho\sigma}}{\partial x^{\mu}}\frac{\sqrt{-g}}{2}g^{\rho\alpha}g^{\sigma\beta}A_{\alpha\beta}$$

Applying (13) in the second term on the right then with change of summation indices α to ρ and β to σ, A_{μ} - the divergence of the mixed tensor - is expressed in:

$$\sqrt{-g}A_{\mu} = \frac{\partial}{\partial x^{\sigma}}\left(\sqrt{-g}A^{\sigma}_{\mu}\right) + \frac{\partial g^{\rho\sigma}}{\partial x^{\mu}}\frac{\sqrt{-g}A_{\rho\sigma}}{2} \tag{28}$$

J. The Metric Tensor III – The Riemann-Christoffel Tensor

Here we seek the tensors that are formed by differentiating the metric tensor.

First Approach

Substitute $g_{\mu\nu}$ for $A_{\mu\nu}$ in equation (19):

$$g_{\mu\nu\sigma} = \frac{\partial g_{\mu\nu}}{\partial x^{\sigma}} - \{\sigma\mu,\tau\}g_{\tau\nu} - \{\sigma\nu,\tau\}g_{\tau\mu}$$

Recalling (6), rewrite the equation above as:

$$\begin{aligned} g_{\mu\nu\sigma} &= \frac{\partial g_{\mu\nu}}{\partial x^{\sigma}} - g^{\tau\alpha}[\sigma\mu,\alpha]g_{\tau\nu} - g^{\tau\alpha}[\sigma\nu,\alpha]g_{\mu\tau} \\ &= \frac{\partial g_{\mu\nu}}{\partial x^{\sigma}} - [\sigma\mu,\nu] - [\sigma\nu,\mu] \end{aligned}$$

as $g^{\tau\alpha}g_{\tau\nu} = 1$ for $\alpha = \nu$. However, substituting equation (14) in the first term on the right shows that $g_{\mu\nu\sigma}$ vanishes. (This outcome is called Ricci's Lemma.)

Second Approach

Instead of $g_{\mu\nu}$, in equation (9) change summation index τ to ρ then substitute it into (19):

$$A_{\mu\nu\sigma} = \frac{\partial}{\partial x^\sigma}\left(\frac{\partial A_\mu}{\partial x^\nu} - \{\mu\nu,\rho\}A_\rho\right) - \{\sigma\mu,\tau\}\left(\frac{\partial A_\tau}{\partial x^\nu} - \{\tau\nu,\rho\}A_\rho\right)$$

$$- \{\sigma\nu,\tau\}\left(\frac{\partial A_\mu}{\partial x^\tau} - \{\mu\tau,\rho\}A_\rho\right)$$

Expanding:

$$A_{\mu\nu\sigma} = \frac{\partial^2 A_\mu}{\partial x^\sigma \partial x^\nu} - A_\rho \frac{\partial}{\partial x^\sigma}\{\mu\nu,\rho\} - \{\mu\nu,\rho\}\frac{\partial A_\rho}{\partial x^\sigma} - \{\sigma\mu,\tau\}\frac{\partial A_\tau}{\partial x^\nu}$$

$$+ \{\sigma\mu,\tau\}\{\tau\nu,\rho\}A_\rho - \{\sigma\nu,\tau\}\frac{\partial A_\mu}{\partial x^\tau} + \{\sigma\nu,\tau\}\{\mu\tau,\rho\}A_\rho$$

With sequential notational changes of τ to α, σ to τ and ν to σ:

$$A_{\mu\sigma\tau} = \frac{\partial^2 A_\mu}{\partial x^\tau \partial x^\sigma} - A_\rho \frac{\partial}{\partial x^\tau}\{\mu\sigma,\rho\} - \{\mu\sigma,\rho\}\frac{\partial A_\rho}{\partial x^\tau} - \{\tau\mu,\alpha\}\frac{\partial A_\alpha}{\partial x^\sigma}$$

$$+ \{\tau\mu,\alpha\}\{\alpha\sigma,\rho\}A_\rho - \{\tau\sigma,\alpha\}\frac{\partial A_\mu}{\partial x^\alpha} + \{\tau\sigma,\alpha\}\{\mu\alpha,\rho\}A_\rho$$

Now interchanging σ and τ:

$$A_{\mu\tau\sigma} = \frac{\partial^2 A_\mu}{\partial x^\sigma \partial x^\tau} - A_\rho \frac{\partial}{\partial x^\sigma}\{\mu\tau,\rho\} - \{\mu\tau,\rho\}\frac{\partial A_\rho}{\partial x^\sigma} - \{\sigma\mu,\alpha\}\frac{\partial A_\alpha}{\partial x^\tau}$$

$$+ \{\sigma\mu,\alpha\}\{\alpha\tau,\rho\}A_\rho - \{\sigma\tau,\alpha\}\frac{\partial A_\mu}{\partial x^\alpha} + \{\sigma\tau,\alpha\}\{\mu\alpha,\rho\}A_\rho$$

Each of the first, sixth and seventh pairs of terms of the preceding two equations are symmetric in τ and σ and cancel in the difference $A_{\mu\sigma\tau} - A_{\mu\tau\sigma}$. In the fourth term of both equations, notational change of summation index α to ρ reveals the symmetry of the sum of the third and fourth terms of both equations in τ and σ. So both sets of terms cancel in the difference of the two equations. Hence:

$$A_{\mu\sigma\tau} - A_{\mu\tau\sigma} = - A_\rho \frac{\partial}{\partial x^\tau}\{\mu\sigma,\rho\} + \{\tau\mu,\alpha\}\{\alpha\sigma,\rho\}A_\rho$$

$$+ A_\rho \frac{\partial}{\partial x^\sigma}\{\mu\tau,\rho\} - \{\sigma\mu,\alpha\}\{\alpha\tau,\rho\}A_\rho$$

$$A_{\mu\sigma\tau} - A_{\mu\tau\sigma} = \left[\frac{\partial}{\partial x^{\sigma}}\{\mu\tau,\rho\} - \frac{\partial}{\partial x^{\tau}}\{\mu\sigma,\rho\} + \{\tau\mu,\alpha\}\{\alpha\sigma,\rho\} - \{\sigma\mu,\alpha\}\{\alpha\tau,\rho\}\right] A_{\rho}$$
$$= \mathfrak{R}^{\rho}_{\mu\sigma\tau}\, A_{\rho}$$

Where: $$\mathfrak{R}^{\rho}_{\mu\sigma\tau} = \frac{\partial}{\partial x^{\sigma}}\{\mu\tau,\rho\} - \frac{\partial}{\partial x^{\tau}}\{\mu\sigma,\rho\} + \{\tau\mu,\alpha\}\{\alpha\sigma,\rho\} - \{\sigma\mu,\alpha\}\{\alpha\tau,\rho\} \qquad (29a)$$

Recalling equation (7), this may be rewritten as:

$$\mathfrak{R}^{\rho}_{\mu\sigma\tau} = -\frac{\partial\Gamma^{\rho}_{\mu\tau}}{\partial x^{\sigma}} + \frac{\partial\Gamma^{\rho}_{\mu\sigma}}{\partial x^{\tau}} + \Gamma^{\alpha}_{\tau\mu}\Gamma^{\rho}_{\alpha\sigma} - \Gamma^{\alpha}_{\sigma\mu}\Gamma^{\rho}_{\alpha\tau} \qquad (29b)$$

Since $A_{\mu\sigma\tau}$, $A_{\mu\tau\sigma}$ and A_{ρ} are tensors then $\mathfrak{R}^{\rho}_{\mu\sigma\tau}$ is also a tensor. It is called the *Riemann-Christoffel tensor*. This tensor contains, as multiplicative factors in every term, coordinate derivatives of $g_{\mu\nu}$. Hence for constant $g_{\mu\nu}$, $\mathfrak{R}^{\rho}_{\mu\sigma\tau}$ vanishes. Due to the homogeneousness of its coordinate transformations, once $\mathfrak{R}^{\rho}_{\mu\sigma\tau}$ vanishes in one chart of a region, it vanishes in all charts of the same region, even those with $g_{\mu\nu}$ that are functions of the coordinates.

K. The Field Equations of Gravity in the Absence of Matter

Matter, in Einstein's use of the term in his *FGTR*, includes all energetic entities and processes apart from gravity. This manner of usage will generally be adhered to in going forward.

The vanishing of $\mathfrak{R}^{\rho}_{\mu\sigma\tau}$ is the necessary and sufficient condition for the application of a Minkowskian chart over a *finite* region.[19] Such gravity-free regions may only exist remote from any matter. In the cases of finite matter-free regions proximate to gravitating bodies, the vanishing of $\mathfrak{R}^{\rho}_{\mu\sigma\tau}$ is an excessive requirement. Yet, in any infinitesimal region a Minkowskian chart may be applied. This may occur along geodesics traversing infinitesimal regions. [cf. § F.] Therefore, for such cases, we must then seek a suitable tensor that may be derived from $\mathfrak{R}^{\rho}_{\mu\sigma\tau}$. Furthermore, it is required that in finite regions remote from any matter, the vanishing of the tensor being sought must yield, in each chart, the ten quantities $g_{\mu\nu}$ – functions of coordinates or constants - required for the vanishing of $\mathfrak{R}^{\rho}_{\mu\sigma\tau}$.

Consider contracting equation (29a) with respect to ρ and τ, let $\rho = \tau = \beta$:

$$\mathfrak{R}_{\mu\sigma} = \mathfrak{R}^{\beta}_{\mu\sigma\beta} = \frac{\partial}{\partial x^{\sigma}}\{\mu\beta,\beta\} - \frac{\partial}{\partial x^{\beta}}\{\mu\sigma,\beta\} + \{\beta\mu,\alpha\}\{\alpha\sigma,\beta\} - \{\sigma\mu,\alpha\}\{\alpha\beta,\beta\} \quad (29c)$$

Substituting from equation (17), changing σ to ν and interchanging β and α yield:

$$\mathfrak{R}_{\mu\nu} = \frac{\partial^2 \ln\sqrt{-g}}{\partial x^{\nu}\partial x^{\mu}} - \frac{\partial}{\partial x^{\alpha}}\{\mu\nu,\alpha\} + \{\alpha\mu,\beta\}\{\beta\nu,\alpha\} - \{\mu\nu,\beta\}\frac{\partial \ln\sqrt{-g}}{\partial x^{\beta}} = R_{\mu\nu} + S_{\mu\nu}$$

where we let: $$R_{\mu\nu} = -\frac{\partial}{\partial x^{\alpha}}\{\mu\nu,\alpha\} + \{\alpha\mu,\beta\}\{\beta\nu,\alpha\} \quad (30a)$$

and, $$S_{\mu\nu} = \frac{\partial^2 \ln\sqrt{-g}}{\partial x^{\nu}\partial x^{\mu}} - \{\mu\nu,\beta\}\frac{\partial \ln\sqrt{-g}}{\partial x^{\beta}} \quad (30b)$$

For $\sqrt{-g} = 1$: $S_{\mu\nu} = 0$ and $\mathcal{R}_{\mu\nu} = R_{\mu\nu}$

$R_{\mu\nu}$ is known as the *Ricci tensor* or the *curvature*.

Recalling (7), equation (30a) may be rewritten as:

$$R_{\mu\nu} = \frac{\partial \Gamma^{\alpha}_{\mu\nu}}{\partial x^{\alpha}} + \Gamma^{\beta}_{\mu\alpha}\Gamma^{\alpha}_{\beta\nu} \quad (31)$$

As $\Gamma^{\alpha}_{\mu\nu}$ is symmetrical in μ and ν so is $R_{\mu\nu}$. This tensor vanishes for constant $g_{\mu\nu}$. Generally, its vanishing yields the *field equation of gravity in the absence of matter*:

$$R_{\mu\nu} = \frac{\partial \Gamma^{\alpha}_{\mu\nu}}{\partial x^{\alpha}} + \Gamma^{\beta}_{\mu\alpha}\Gamma^{\alpha}_{\beta\nu} = 0 \qquad : \sqrt{-g} = 1 \quad (32)$$

Einstein gave the following reasons for this assertion. $\mathcal{R}_{\mu\nu}$ is the only tensor, formed from the $g_{\mu\nu}$ alone, that:

a. is of the second rank (yielding the ten quantities $g_{\mu\nu}$)
b. contains no derivatives of $g_{\mu\nu}$ that is higher than the second
c. is linear in the second derivatives.

In addition, equations (8) and (32) (specifically the latter's inhomogenous form that will be discussed in due course):

d. to a first approximation yields Newtonian laws of gravity, and
e. to a second approximation yields an explanation for the previously curious advance of the perihelion (point of nearest approach to the sun) of Mercury.

L. Energy and Momentum of the Gravitational Field in the Absence of Matter

Assume the Hamiltonian function:

$$H\left(g^{\mu\nu}(x^{\sigma}), \frac{\partial g^{\mu\nu}(x^{\sigma})}{\partial x^{\sigma}}\right) = g^{\mu\nu}\Gamma^{\alpha}_{\mu\beta}\Gamma^{\beta}_{\nu\alpha} \qquad : \sqrt{-g} = 1 \quad (33)$$

is the energy of the matter-free gravitational field. To determine the relations of energy and momentum in the matter-free field, perform the following variation:

$$\delta \int_V H\, d\tau = \int_V \delta H\, d\tau = 0 \quad (34)$$

Here $d\tau$ is the differential 4-volume element and V is a finite 4-D volume. The variations vanish on the boundary of the finite element of integration.

Varying (33) yields:

$$\delta H = \Gamma^{\alpha}_{\mu\beta}\Gamma^{\beta}_{\nu\alpha}\delta g^{\mu\nu} + 2g^{\mu\nu}\Gamma^{\alpha}_{\mu\beta}\delta\Gamma^{\beta}_{\nu\alpha}$$

Now:

$$g^{\mu\nu}\delta\Gamma^{\beta}_{\nu\alpha} = \delta\left(g^{\mu\nu}\Gamma^{\beta}_{\nu\alpha}\right) - \Gamma^{\beta}_{\nu\alpha}\delta g^{\mu\nu}$$

So:

$$\delta H = \Gamma^{\alpha}_{\mu\beta}\Gamma^{\beta}_{\nu\alpha}\delta g^{\mu\nu} + 2\Gamma^{\alpha}_{\mu\beta}\left[\delta\left(g^{\mu\nu}\Gamma^{\beta}_{\nu\alpha}\right) - \Gamma^{\beta}_{\nu\alpha}\delta g^{\mu\nu}\right]$$

$$= -\Gamma^{\alpha}_{\mu\beta}\Gamma^{\beta}_{\nu\alpha}\delta g^{\mu\nu} + 2\Gamma^{\alpha}_{\mu\beta}\left[\delta\left(g^{\mu\nu}\Gamma^{\beta}_{\nu\alpha}\right)\right]$$

Also, recalling equation (7):

$$\Gamma^{\alpha}_{\mu\beta}\left[\delta\left(g^{\mu\nu}\Gamma^{\beta}_{\nu\alpha}\right)\right] = -\frac{1}{2}\Gamma^{\alpha}_{\mu\beta}\delta\left[g^{\mu\nu}g^{\beta\lambda}\left(\frac{\partial g_{\nu\lambda}}{\partial x^{\alpha}} + \frac{\partial g_{\alpha\lambda}}{\partial x^{\nu}} - \frac{\partial g_{\nu\alpha}}{\partial x^{\lambda}}\right)\right]$$

The difference of the last two terms in the round brackets is anti-symmetric in ν and λ, multiplied by $g^{\beta\nu}g^{\mu\lambda}$, it becomes anti-symmetric in μ and β. Therefore, in light of the discussion in the first paragraph of § I.3.2.1, multiplication by $\Gamma^{\alpha}_{\mu\beta}$, which is symmetric in the same indices, causes mutual cancellation. Therefore applying (13), the equation above becomes:

$$\Gamma^{\alpha}_{\mu\beta}\left[\delta\left(g^{\mu\nu}\Gamma^{\beta}_{\nu\alpha}\right)\right] = -\frac{\Gamma^{\alpha}_{\mu\beta}}{2}\delta\left(g^{\mu\nu}g^{\beta\lambda}\frac{\partial g_{\nu\lambda}}{\partial x^{\alpha}}\right) = \frac{1}{2}\Gamma^{\alpha}_{\mu\beta}\delta g^{\mu\beta}_{\alpha}$$

So

$$\delta H = -\Gamma^{\alpha}_{\mu\beta}\Gamma^{\beta}_{\nu\alpha}\delta g^{\mu\nu} + \Gamma^{\alpha}_{\mu\beta}\delta g^{\mu\beta}_{\alpha} = -\Gamma^{\alpha}_{\mu\beta}\Gamma^{\beta}_{\nu\alpha}\delta g^{\mu\nu} + \Gamma^{\sigma}_{\mu\nu}\delta g^{\mu\nu}_{\sigma}$$

after changing summation indices α to σ and β to ν in the last term.

Now varying H with respect to the canonical coordinates - $g^{\mu\nu}$ and $g^{\mu\nu}_{\sigma}$ - yields:

$$\delta H(g^{\mu\nu}, g^{\mu\nu}_{\sigma}) = \delta g^{\mu\nu}\frac{\partial H}{\partial g^{\mu\nu}} + \delta g^{\mu\nu}_{\sigma}\frac{\partial H}{\partial g^{\mu\nu}_{\sigma}}$$

Equating coefficients of the variations in the last two equations we get:

$$\frac{\partial H}{\partial g^{\mu\nu}} = -\Gamma^{\alpha}_{\mu\beta}\Gamma^{\beta}_{\nu\alpha} = -\Gamma^{\beta}_{\mu\alpha}\Gamma^{\alpha}_{\nu\beta} \text{ and } \frac{\partial H}{\partial g^{\mu\nu}_{\sigma}} = \Gamma^{\sigma}_{\mu\nu} \text{ or } \frac{\partial H}{\partial g^{\mu\nu}_{\alpha}} = \Gamma^{\alpha}_{\mu\nu} \qquad (35)$$

after interchange of summation indices in the first equation and change of non-summation index σ to α in the last equation.

Performing the variation of (34) with respect to the coordinates - x^{α}:

$$\int_V \delta H d\tau = \int_V \frac{\partial H}{\partial x^\alpha}\delta x^\alpha d\tau = \int_V \left[\frac{\partial H}{\partial g^{\mu\nu}}\frac{\partial g^{\mu\nu}}{\partial x^\alpha} + \frac{\partial H}{\partial g^{\mu\nu}_\alpha}\frac{\partial g^{\mu\nu}_\alpha}{\partial x^\alpha}\right]\delta x^\alpha d\tau$$

Rewriting the second term in the brackets:

$$\int_V \frac{\partial H}{\partial x^\alpha}\delta x^\alpha d\tau = \int_V \left[\frac{\partial H}{\partial g^{\mu\nu}} g^{\mu\nu}_\alpha + \frac{\partial}{\partial x^\alpha}\left(g^{\mu\nu}_\alpha \frac{\partial H}{\partial g^{\mu\nu}_\alpha}\right) - g^{\mu\nu}_\alpha \frac{\partial}{\partial x^\alpha}\left(\frac{\partial H}{\partial g^{\mu\nu}_\alpha}\right)\right]\delta x^\alpha d\tau$$

Partially integrate the second term of the expanded right side above:

$$\int_V \left[\frac{\partial}{\partial x^\alpha}\left(g^{\mu\nu}_\alpha \frac{\partial H}{\partial g^{\mu\nu}_\alpha}\right)\right]\delta x^\alpha d\tau = \left[g^{\mu\nu}_\alpha \frac{\partial H}{\partial g^{\mu\nu}_\alpha}\delta x^\alpha\right]\Bigg|_{\tau_i}^{\tau_f} - \int_V g^{\mu\nu}_\alpha \frac{\partial H}{\partial g^{\mu\nu}_\alpha}\frac{\partial \delta x^\alpha}{\partial x^\alpha} d\tau$$

where τ_i and τ_f represent, respectively, initial and final limits of integration at which δx^α vanishes. Furthermore,

$$\frac{\partial \delta x^\alpha}{\partial x^\alpha} = \delta\left(\frac{\partial x^\alpha}{\partial x^\alpha}\right) = \delta(4) = 0$$

So:
$$\int_V \frac{\partial H}{\partial x^\alpha}\delta x^\alpha d\tau = \int_V \left[\frac{\partial H}{\partial g^{\mu\nu}} - \frac{\partial}{\partial x^\alpha}\left(\frac{\partial H}{\partial g^{\mu\nu}_\alpha}\right)\right] g^{\mu\nu}_\alpha \delta x^\alpha d\tau$$

Then, for arbitrary values of the product $g^{\mu\nu}_\alpha \delta x^\alpha$, stationarity requires:

$$\frac{\partial}{\partial x^\alpha}\left(\frac{\partial H}{\partial g^{\mu\nu}_\alpha}\right) - \frac{\partial H}{\partial g^{\mu\nu}} = 0$$

Substituting from (35), this last equation yields (32). Hence the assumption of (33) is validated. From which it follows that the same applies for the inference that the quantity $\Gamma^\tau_{\mu\nu}$ introduced in equation (8) is the field component of gravity.

Multiplying the last equation above by $g^{\mu\nu}_\sigma$ and rewriting:

$$g^{\mu\nu}_\sigma \frac{\partial}{\partial x^\alpha}\left(\frac{\partial H}{\partial g^{\mu\nu}_\alpha}\right) - g^{\mu\nu}_\sigma \frac{\partial H}{\partial g^{\mu\nu}} = \frac{\partial}{\partial x^\alpha}\left(g^{\mu\nu}_\sigma \frac{\partial H}{\partial g^{\mu\nu}_\alpha}\right) - \frac{\partial g^{\mu\nu}_\sigma}{\partial x^\alpha}\frac{\partial H}{\partial g^{\mu\nu}_\alpha} - \frac{\partial g^{\mu\nu}}{\partial x^\sigma}\frac{\partial H}{\partial g^{\mu\nu}} = 0$$

The commutativity of the partial differentiation implies:

$$\frac{\partial g^{\mu\nu}_\sigma}{\partial x^\alpha} = \frac{\partial}{\partial x^\alpha}\left(\frac{\partial g^{\mu\nu}}{\partial x^\sigma}\right) = \frac{\partial}{\partial x^\sigma}\left(\frac{\partial g^{\mu\nu}}{\partial x^\alpha}\right) = \frac{\partial g^{\mu\nu}_\alpha}{\partial x^\sigma}$$

Therefore, the previous equation becomes:

$$\frac{\partial}{\partial x^\alpha}\left(g_\sigma^{\mu\nu}\frac{\partial H}{\partial g_\alpha^{\mu\nu}}\right)-\frac{\partial g_\alpha^{\mu\nu}}{\partial x^\sigma}\frac{\partial H}{\partial g_\alpha^{\mu\nu}}-\frac{\partial g^{\mu\nu}}{\partial x^\sigma}\frac{\partial H}{\partial g^{\mu\nu}}=\frac{\partial}{\partial x^\alpha}\left(g_\sigma^{\mu\nu}\frac{\partial H}{\partial g_\alpha^{\mu\nu}}\right)-\frac{\partial H}{\partial x^\sigma}=0$$

Now:

$$\frac{\partial}{\partial x^\alpha}(\delta_\sigma^\alpha H)=\frac{\partial H}{\partial x^\sigma} \quad : \alpha=\sigma \tag{36a}$$
$$=0 \quad : \alpha\neq\sigma$$

Therefore we may write:

$$\frac{\partial}{\partial x^\alpha}\left(\delta_\sigma^\alpha H - g_\sigma^{\mu\nu}\frac{\partial H}{\partial g_\alpha^{\mu\nu}}\right)=\frac{\partial t_\sigma^\alpha}{\partial x^\alpha}=0 \tag{36b}$$

$$\delta_\sigma^\alpha H - g_\sigma^{\mu\nu}\frac{\partial H}{\partial g_\alpha^{\mu\nu}} = 2\kappa t_\sigma^\alpha \tag{37}$$

where 2κ is a constant of proportionality (to be determined).

Changing the notation of the summation index α to λ in (33) and recalling (35) and (16), then substituting in (37):

$$2\kappa t_\sigma^\alpha = \delta_\sigma^\alpha g^{\mu\nu}\Gamma_{\mu\beta}^\lambda\Gamma_{\nu\lambda}^\beta - (g^{\mu\tau}\Gamma_{\tau\sigma}^\nu + g^{\nu\tau}\Gamma_{\tau\sigma}^\mu)\Gamma_{\mu\nu}^\alpha$$
$$= \delta_\sigma^\alpha g^{\mu\nu}\Gamma_{\mu\beta}^\lambda\Gamma_{\nu\lambda}^\beta - g^{\mu\tau}\Gamma_{\tau\sigma}^\nu\Gamma_{\mu\nu}^\alpha - g^{\nu\tau}\Gamma_{\tau\sigma}^\mu\Gamma_{\mu\nu}^\alpha$$

Changing the notation of the summation indices ν to β and τ to ν in the second term and, in the third term, μ to β, ν to μ and τ to ν reveal their identity, therefore:

$$\kappa t_\sigma^\alpha = \delta_\sigma^\alpha g^{\mu\nu}\Gamma_{\mu\beta}^\lambda\Gamma_{\nu\lambda}^\beta/2 - g^{\mu\nu}\Gamma_{\mu\beta}^\alpha\Gamma_{\nu\sigma}^\beta \tag{38}$$

This equation describes the relationship of conservation of energy and momentum of the matter-free gravitational field. The t_σ^α are the energy and momentum components of the matter-free gravitational field. They vanish in a frame of reference that moves along a geodesic world-line (4-D locus): observers moving with such a frame would not detect any gravitational energy - components of which are the t_σ^α [cf. § F.]. So, unlike tensors, under a transformation to a chart that moves on a geodesic world-line, a formerly non-zero t_σ^α would vanish. The t_σ^α are *pseudo-tensors* as though not being tensors - as we shall see - they may, in some ways, be treated as tensors.

The vanishing of $\partial t_\sigma^\alpha/\partial x_\alpha$ in equation (36b) is the expression of the conservation of energy and momentum of the field within a region free of matter.

Multiply equation (32) by the contravariant metric tensor:

$$g^{\nu\sigma}\frac{\partial\Gamma_{\mu\nu}^\alpha}{\partial x^\alpha} + g^{\nu\sigma}\Gamma_{\mu\alpha}^\beta\Gamma_{\beta\nu}^\alpha = 0$$

Changing the summation index ν to β in the first term and rewriting:

$$\frac{\partial}{\partial x^\alpha}\left(g^{\beta\sigma}\Gamma^\alpha_{\mu\beta}\right) - \Gamma^\alpha_{\mu\beta}\frac{\partial g^{\beta\sigma}}{\partial x^\alpha} + g^{\nu\sigma}\Gamma^\beta_{\mu\alpha}\Gamma^\alpha_{\beta\nu} = 0$$

Recalling (16): $\frac{\partial}{\partial x^\alpha}\left(g^{\beta\sigma}\Gamma^\alpha_{\mu\beta}\right) - g^{\beta\tau}\Gamma^\sigma_{\tau\alpha}\Gamma^\alpha_{\mu\beta} - g^{\sigma\tau}\Gamma^\beta_{\tau\alpha}\Gamma^\alpha_{\mu\beta} + g^{\nu\sigma}\Gamma^\beta_{\mu\alpha}\Gamma^\alpha_{\beta\nu} = 0$

In the third term, interchanging the summation indices α and β and changing the summation index τ to ν reveal the cancellation of this term with the fourth term.

And $$\frac{\partial}{\partial x^\alpha}\left(g^{\beta\sigma}\Gamma^\alpha_{\mu\beta}\right) = g^{\beta\tau}\Gamma^\sigma_{\tau\alpha}\Gamma^\alpha_{\mu\beta} \qquad : \sqrt{-g} = 1 \qquad (39)$$

Contracting equation (38) with respect to σ and α, let $\sigma = \alpha = \lambda$ and recall that $\delta^\lambda_\lambda = 4$, then:

$$\kappa t^\lambda_\lambda = \kappa t = 2g^{\mu\nu}\Gamma^\lambda_{\mu\beta}\Gamma^\beta_{\nu\lambda} - g^{\mu\nu}\Gamma^\lambda_{\mu\beta}\Gamma^\beta_{\nu\lambda} = g^{\mu\nu}\Gamma^\lambda_{\mu\beta}\Gamma^\beta_{\nu\lambda}$$

Multiplying the above equation by $\delta^\alpha_\sigma/2$ and subtracting from (38):

$$\kappa t^\alpha_\sigma - \delta^\alpha_\sigma \kappa t/2 = -\, g^{\mu\nu}\Gamma^\alpha_{\mu\beta}\Gamma^\beta_{\nu\sigma}$$

Then, sequentially change the notations of the indices as follows: μ to τ, σ to μ, α to σ, β to α and ν to β. This yields:

$$\kappa t^\sigma_\mu - \delta^\sigma_\mu \kappa t/2 = -\, g^{\tau\beta}\Gamma^\sigma_{\tau\alpha}\Gamma^\alpha_{\beta\mu}$$

Substituting in (39): $$\frac{\partial}{\partial x^\alpha}\left(g^{\beta\sigma}\Gamma^\alpha_{\mu\beta}\right) = -\kappa\left(t^\sigma_\mu - \frac{\delta^\sigma_\mu t}{2}\right) \qquad : \sqrt{-g} = 1 \qquad (40)$$

Equations (39) and (40) are *alternative forms of the matter-free field equation*. Equation (40) explicitly expresses the relationship of the field components $\Gamma^\alpha_{\mu\beta}$ and the energy and momentum components t^σ_μ of gravity in a matter-free region.

M. The General Field Equations of Gravity - Including Matter

STR presents material energy in a symmetrical tensor of the second rank - T^j_k. *Assume* that the gravitational field is induced by the energy and momentum components of both the gravitational and material fields combined as $t^\sigma_\mu + T^\sigma_\mu$. As such, in equation (40), replace t^σ_μ with this sum and likewise replace t with the sum t + T. The resulting *general field equations of gravity* for $\sqrt{-g} = 1$ are:

$$\frac{\partial}{\partial x^\alpha}\left(g^{\beta\sigma}\Gamma^\alpha_{\mu\beta}\right) = -\kappa\left(t^\sigma_\mu + T^\sigma_\mu\right) + \frac{\kappa\delta^\sigma_\mu}{2}(t + T) \quad : \sigma, \mu = 1,2,3,4 \; (41)$$

These substitutions amount to an assertion that the *energy and momentum of gravity act gravitatively as do those of matter*. However, since the gravitational field is itself induced by matter, we seek to express its field equations in terms of the energy and momentum of matter. Rewriting equation (41) as follows:

$$\frac{\partial}{\partial x^\alpha}\left(g^{\beta\sigma}\Gamma^\alpha_{\mu\beta}\right) = -\kappa\left(t^\sigma_\mu - \frac{\delta^\sigma_\mu t}{2}\right) - \kappa\left(T^\sigma_\mu - \frac{\delta^\sigma_\mu T}{2}\right)$$

Now utilizing equations (39) and (40) to eliminate the first term on the right and then differentiating:

$$g^{\beta\sigma}_\alpha\Gamma^\alpha_{\mu\beta} + g^{\beta\sigma}\frac{\partial\Gamma^\alpha_{\mu\beta}}{\partial x^\alpha} = g^{\beta\tau}\Gamma^\sigma_{\tau\alpha}\Gamma^\alpha_{\mu\beta} - \kappa\left(T^\sigma_\mu - \frac{\delta^\sigma_\mu T}{2}\right)$$

Recalling equation (16) rewrite the first term above:

$$g^{\beta\tau}\Gamma^\sigma_{\tau\alpha}\Gamma^\alpha_{\mu\beta} + g^{\sigma\tau}\Gamma^\beta_{\tau\alpha}\Gamma^\alpha_{\mu\beta} + g^{\beta\sigma}\frac{\partial\Gamma^\alpha_{\mu\beta}}{\partial x^\alpha} = g^{\beta\tau}\Gamma^\sigma_{\tau\alpha}\Gamma^\alpha_{\mu\beta} - \kappa\left(T^\sigma_\mu - \frac{\delta^\sigma_\mu T}{2}\right)$$

and we get:

$$g^{\beta\sigma}\frac{\partial\Gamma^\alpha_{\mu\beta}}{\partial x^\alpha} + g^{\sigma\tau}\Gamma^\beta_{\tau\alpha}\Gamma^\alpha_{\mu\beta} = -\kappa\left(T^\sigma_\mu - \frac{\delta^\sigma_\mu T}{2}\right)$$

On the left, interchange summation indices α and β of the second term, multiply the equation by $g_{\sigma\nu}$, then contract the terms on the left and lower indices on the right (recalling that $\delta^\sigma_\mu = g^\sigma_\mu$) yields the inhomogenous form of equation (32):

$$\frac{\partial\Gamma^\alpha_{\mu\nu}}{\partial x^\alpha} + \Gamma^\beta_{\mu\alpha}\Gamma^\alpha_{\beta\nu} = -\kappa\left(T_{\mu\nu} - \frac{g_{\mu\nu}T}{2}\right) \qquad : \sqrt{-g} = 1 \qquad (42)$$

Recalling (31):

$$R_{\mu\nu} = -\kappa(T_{\mu\nu} - g_{\mu\nu}T/2) \qquad (43)$$

Metrically contracting this last equation by $g^{\mu\nu}$ yields:

$$R = -\kappa(T - 4T/2) = \kappa T \qquad (44)$$

and (43) and (44) yield:

$$R_{\mu\nu} - g_{\mu\nu}R/2 = -\kappa T_{\mu\nu} \qquad (45)$$

The four preceding equations are forms of the *general field equations of gravity*. In mathematics, R is called the *scalar curvature*.

N. Conservation of Energy and Momentum in the General Case

Contracting equation (41) with respect to μ and σ, let $\mu = \sigma = \lambda$:

$$\frac{\partial}{\partial x^\alpha}\left(g^{\lambda\beta}\Gamma^\alpha_{\lambda\beta}\right) = -\kappa(t + T) + \frac{4\kappa}{2}(t + T) = \kappa(t + T)$$

Multiply this equation by $\delta^\sigma_\mu/2$ and subtract it from equation (41), then differentiate with respect to the coordinates - x^σ:

$$\frac{\partial}{\partial x^\sigma\partial x^\alpha}\left(g^{\sigma\beta}\Gamma^\alpha_{\mu\beta} - \frac{\delta^\sigma_\mu}{2}g^{\lambda\beta}\Gamma^\alpha_{\lambda\beta}\right) = \frac{\partial}{\partial x^\sigma}\left[-\kappa(t^\sigma_\mu + T^\sigma_\mu)\right] \qquad (46a)$$

Apply (7) to the first term of the expanded left side:

$$\frac{\partial^2}{\partial x^\sigma \partial x^\alpha}(g^{\sigma\beta}\Gamma^\alpha_{\mu\beta}) = -\frac{1}{2}\frac{\partial^2}{\partial x^\sigma \partial x^\alpha}\left[g^{\sigma\beta}g^{\alpha\lambda}\left(\frac{\partial g_{\mu\lambda}}{\partial x^\beta} + \frac{\partial g_{\beta\lambda}}{\partial x^\mu} - \frac{\partial g_{\mu\beta}}{\partial x^\lambda}\right)\right] \tag{46b}$$

Consider:

$$\frac{\partial^2}{\partial x^\sigma \partial x^\alpha}\left[g^{\sigma\beta}g^{\alpha\lambda}\left(\frac{\partial g_{\mu\lambda}}{\partial x^\beta} - \frac{\partial g_{\mu\beta}}{\partial x^\lambda}\right)\right] = \frac{\partial^2}{\partial x^\sigma \partial x^\alpha}\left(g^{\sigma\beta}g^{\alpha\lambda}\frac{\partial g_{\mu\lambda}}{\partial x^\beta}\right) - \frac{\partial^2}{\partial x^\sigma \partial x^\alpha}\left(g^{\sigma\beta}g^{\alpha\lambda}\frac{\partial g_{\mu\beta}}{\partial x^\lambda}\right)$$

In the first term on the right, interchange the summation indices λ and β on one hand, and on the other, the summation indices σ and α. These changes, with commutativity of the partial differentiation, produce an expression that is the negative of the second term. Therefore, in summation, they cancel. Now, invoke equation (13) on the remaining term of the expanded right side of equation (46b) and it becomes:

$$\frac{\partial^2}{\partial x^\sigma \partial x^\alpha}(g^{\sigma\beta}\Gamma^\alpha_{\mu\beta}) = -\frac{1}{2}\frac{\partial^2}{\partial x^\sigma \partial x^\alpha}\left(g^{\sigma\beta}g^{\alpha\lambda}\frac{\partial g_{\beta\lambda}}{\partial x^\mu}\right) = \frac{1}{2}\frac{\partial^3 g^{\sigma\alpha}}{\partial x^\sigma \partial x^\alpha \partial x^\mu} \tag{46c}$$

Next, apply (7) to the second term of the expanded left side of (46a):

$$-\frac{\partial^2}{\partial x^\sigma \partial x^\alpha}\left(\frac{1}{2}\delta^\sigma_\mu g^{\lambda\beta}\Gamma^\alpha_{\lambda\beta}\right) = \frac{\partial^2}{\partial x^\sigma \partial x^\alpha}\left[\frac{\delta^\sigma_\mu}{4}g^{\lambda\beta}g^{\alpha\tau}\left(\frac{\partial g_{\tau\lambda}}{\partial x^\beta} + \frac{\partial g_{\tau\beta}}{\partial x^\lambda} - \frac{\partial g_{\lambda\beta}}{\partial x^\tau}\right)\right] \tag{46d}$$

For $\sqrt{-g} = 1$, equation (11a) implies the vanishing of the third term of the expanded right side of (46d). Now, consider:

$$g^{\lambda\beta}g^{\alpha\tau}\left(\frac{\partial g_{\tau\lambda}}{\partial x^\beta} + \frac{\partial g_{\tau\beta}}{\partial x^\lambda}\right) = g^{\lambda\beta}g^{\alpha\tau}\frac{\partial g_{\tau\lambda}}{\partial x^\beta} + g^{\lambda\beta}g^{\alpha\tau}\frac{\partial g_{\tau\beta}}{\partial x^\lambda}$$

On the right, interchange of summation indices λ and β in the second term reveals its equality with the first. Applying equation (12) yields:

$$g^{\lambda\beta}g^{\alpha\tau}\left(\frac{\partial g_{\tau\lambda}}{\partial x^\beta} + \frac{\partial g_{\tau\beta}}{\partial x^\lambda}\right) = 2g^{\lambda\beta}g^{\alpha\tau}\frac{\partial g_{\tau\lambda}}{\partial x^\beta} = -2\frac{\partial g^{\alpha\beta}}{\partial x^\beta}$$

With this result and as $\delta^\sigma_\mu = 1$ for $\sigma = \mu$, we may write (46d) as:

$$\frac{\partial^2}{\partial x^\sigma \partial x^\alpha}\left(\frac{1}{2}\delta^\sigma_\mu g^{\lambda\beta}\Gamma^\alpha_{\lambda\beta}\right) = \frac{\partial^2}{\partial x^\sigma \partial x^\alpha}\left[\frac{\delta^\sigma_\mu}{2}\left(\frac{\partial g^{\beta\alpha}}{\partial x^\beta}\right)\right] = \frac{\partial^2}{\partial x^\mu \partial x^\alpha}\left[\frac{1}{2}\left(\frac{\partial g^{\beta\alpha}}{\partial x^\beta}\right)\right]$$

Changing summation index β to σ in the last term, we may rewrite this as:

$$\frac{\partial^2}{\partial x^\sigma \partial x^\alpha}\left(\frac{1}{2}\delta^\sigma_\mu g^{\lambda\beta}\Gamma^\alpha_{\lambda\beta}\right) = \frac{1}{2}\frac{\partial^3 g^{\sigma\alpha}}{\partial x^\mu \partial x^\alpha \partial x^\sigma} \qquad (46e)$$

The commutativity of the partial differentiation reveals the identity of the expressions on the right of both equations (46c) and (46e), so (46a) becomes:

$$\frac{\partial^2}{\partial x^\sigma \partial x^\alpha}\left(g^{\sigma\beta}\Gamma^\alpha_{\mu\beta} - \frac{\delta^\sigma_\mu}{2} g^{\lambda\beta}\Gamma^\alpha_{\lambda\beta}\right) = 0 = \frac{\partial}{\partial x^\sigma}[-\kappa(t^\sigma_\mu + T^\sigma_\mu)]$$

Or: $$\frac{\partial}{\partial x^\sigma}(t^\sigma_\mu + T^\sigma_\mu) = 0 \qquad : \sqrt{-g} = 1 \quad (47)$$

This equation expresses the conservation of energy and momentum in the general case. Einstein regarded this outcome as justification of the assumptions that the total energy of an isolated system results from both its gravitational and material energies and that the induced gravitational field results from these energies combined simply as the sum of their individual components.

O. Momentum and Energy in Matter/Gravity Interactions

Substitute equations (35) into (42) and multiply by $g^{\mu\nu}_\sigma$:

$$g^{\mu\nu}_\sigma\left[\frac{\partial}{\partial x^\alpha}\left(\frac{\partial H}{\partial g^{\mu\nu}_\alpha}\right) - \frac{\partial H}{\partial g^{\mu\nu}}\right] = -\kappa g^{\mu\nu}_\sigma\left(T_{\mu\nu} - \frac{g_{\mu\nu}T}{2}\right)$$

Recalling (11a), for $\sqrt{-g} = 1$, the second term of the expanded right side vanishes, so:

$$g^{\mu\nu}_\sigma \frac{\partial}{\partial x^\alpha}\left(\frac{\partial H}{\partial g^{\mu\nu}_\alpha}\right) - g^{\mu\nu}_\sigma \frac{\partial H}{\partial g^{\mu\nu}} = -\kappa g^{\mu\nu}_\sigma T_{\mu\nu} \qquad : \sqrt{-g} = 1$$

Rewriting the first term and rearranging the equation yield:

$$g^{\mu\nu}_\sigma \frac{\partial H}{\partial g^{\mu\nu}} - \frac{\partial}{\partial x^\alpha}\left(g^{\mu\nu}_\sigma \frac{\partial H}{\partial g^{\mu\nu}_\alpha}\right) + \frac{\partial H}{\partial g^{\mu\nu}_\alpha}\frac{\partial g^{\mu\nu}_\sigma}{\partial x^\alpha} = \kappa g^{\mu\nu}_\sigma T_{\mu\nu} \qquad (42a)$$

Recalling (36a), we may write:

$$\frac{\partial}{\partial x^\alpha}[\delta^\alpha_\sigma H(g^{\mu\nu}, g^{\mu\nu}_\sigma)] = \frac{\partial}{\partial x^\sigma}[H(g^{\mu\nu}, g^{\mu\nu}_\alpha)] = \frac{\partial H}{\partial g^{\mu\nu}} g^{\mu\nu}_\sigma + \frac{\partial H}{\partial g^{\mu\nu}_\alpha}\frac{\partial g^{\mu\nu}_\alpha}{\partial x^\sigma}$$

Recalling that: $$\frac{\partial g^{\mu\nu}_\alpha}{\partial x^\sigma} = \frac{\partial g^{\mu\nu}_\sigma}{\partial x^\alpha}$$

we may write: $$\frac{\partial}{\partial x^\alpha}(\delta_\sigma^\alpha H) = \frac{\partial H}{\partial g^{\mu\nu}} g_\sigma^{\mu\nu} + \frac{\partial H}{\partial g_\alpha^{\mu\nu}} \frac{\partial g_\sigma^{\mu\nu}}{\partial x^\alpha}$$

Substitute in (42a): $$\frac{\partial}{\partial x^\alpha}(\delta_\sigma^\alpha H) - \frac{\partial}{\partial x^\alpha}\left(g_\sigma^{\mu\nu} \frac{\partial H}{\partial g_\alpha^{\mu\nu}}\right) = \kappa g_\sigma^{\mu\nu} T_{\mu\nu}$$

Being induced by the energy of the material processes, the energy components of the gravitational field t_σ^α may not be conserved. So differentiating equation (37) with respect to x^α and substituting in the above:

$$2\kappa \frac{\partial t_\sigma^\alpha}{\partial x^\alpha} = \frac{\partial}{\partial x^\alpha}(\delta_\sigma^\alpha H) - \frac{\partial}{\partial x^\alpha}\left(g_\sigma^{\mu\nu} \frac{\partial H}{\partial g_\alpha^{\mu\nu}}\right) = \kappa g_\sigma^{\mu\nu} T_{\mu\nu}$$

Or: $$\frac{\partial t_\sigma^\alpha}{\partial x^\alpha} = \frac{1}{2} g_\sigma^{\mu\nu} T_{\mu\nu}$$

Substitute from (47): $$\frac{\partial T_\sigma^\alpha}{\partial x^\alpha} + \frac{1}{2} g_\sigma^{\mu\nu} T_{\mu\nu} = 0 \qquad : \sqrt{-g} = 1$$

As $\sqrt{-g} = 1$, this recalls equation (28) which is one form of the divergence of the mixed tensor of the second rank. Hence:

$$T_\sigma = \frac{\partial T_\sigma^\alpha}{\partial x^\alpha} + \frac{1}{2} g_\sigma^{\mu\nu} T_{\mu\nu} = 0 \qquad : \sqrt{-g} = 1 \qquad (28a)$$

is the *vanishing divergence of the energy tensor*.

Recalling equation (26) for the divergence of a mixed tensor of the second rank and substituting from equations (7) and (28a), we may also write:

$$\frac{\partial T_\mu^\sigma}{\partial x^\sigma} = -\Gamma_{\sigma\mu}^\tau T_\tau^\sigma \qquad : \sqrt{-g} = 1 \qquad (26a)$$

The $\Gamma_{\sigma\mu}^\tau$ are components of the gravitational field. Equations (28a) and (26a) express the interactive dynamics of material and gravitational processes. "Thus the field equations of gravitation contain four conditions (for μ = 1, 2, 3, 4 as the other indices disappear in the summations) which govern the course of material phenomena. They give the equations of material phenomena completely, if the latter is capable of being characterized by four differential equations independent of one another".[20]

Note that the conservation of energy and momentum of material processes does not apply unless the $g^{\mu\nu}$ are constant so that the components of the gravitational field vanish. This is already evident in equation (47).

P. Energy Tensors

According to equation (28a), energy tensors are the components of divergence-free relativistic energy density fluxes of material processes and bodies. As $R_{\mu\nu}$ and $g_{\mu\nu}$ are symmetric in μ and ν, equation (42) implies that $T_{\mu\nu}$ is also. They are functions of the coordinates and often of the proper transitional measurand - ds, the latter as an independent variable with respect to which the coordinates vary. This corresponds to the observer taking measurements relative to his local clock.

The tensors T^{σ}_{σ} for σ =1, 2, 3 are components of energy density fluxes that are *normal stresses* (uni-directional pressures, so to speak) along translational x^{σ}-axes for σ =1, 2, 3, respectively. For example, the 'pressure' T^{1}_{1} acts along the x^1-axis through the infinitesimal surface element of area dx^2dx^3 causing a displacement of dx^1 in a duration of ds. Thus performing work, the normal stress transports energy through the 4-D volume $dx^1dx^2dx^3ds$.

The tensors T^{σ}_{τ}, for τ and σ = 1, 2, 3 and $\sigma \neq \tau$, are relativistic energy density flux components with actions that are transverse to the direction of their flows. These include *shear stresses* and *heat fluxes*. For example, T^{1}_{3} is an energetic flux parallel to, while acting on, the infinitesimal surface element dx^1dx^3. For each pair of values of τ and σ, the flux T^{σ}_{τ} is identical to T^{τ}_{σ}.

The T^{σ}_{4} for σ = 1, 2, 3 are components, each along its corresponding x^{σ}-axis, of the relativistic momentum density due to the translation of relativistic energy density. For example, as material of relativistic energy density ρ transitions at a relative rate of dx^4/ds, while translating at a velocity of dx^1/ds, it produces relativistic momentum density T^{1}_{4} along the x^1-axis. T^{σ}_{4} and T^{4}_{σ} are identical.

The tensor T^{4}_{4} is the transitional flow (rate of change) of relativistic energy within an infinitesimal volume $dx^1dx^2dx^3$ at a relative transitional rate of dx^4/ds.

The material phenomenon that gives rise to these energetic fluxes may be of any type or combination of types including hydrodynamical, electromagnetic and radiative. Normal and shear stresses, momentum, heat and the transition of relativistic energy density are general dynamical aspects of diverse material phenomena.

The general field equations of gravitation, given in equation (42), describe the *induction* of the gravitational field components - $\Gamma^{\tau}_{\mu\nu}$ - *by* the components of the relativistic energy density fluxes of material processes - $T_{\mu\nu}$.

Q. Relativising Physical Relationships

The following are steps that may be used for setting up generally covariant physical relationships:

1. Define the energy tensor T_μ^σ.
2. Insert T_μ^σ in equation (26a). This step produces the generally covariant formulation of the relationships of the physical phenomenon that includes its energetic interaction with the gravitational field.
3. If the $g_{\mu\nu}$ are unknown, then invoke the general field equations (42). The $g_{\mu\nu}$, due to the symmetry in μ and ν, are ten independent functions. Due to the symmetry of $R_{\mu\nu}$, the general field equations yield ten equations for the $g_{\mu\nu}$.
4. The four equations in paragraph 2 are not independent of the general field equations invoked in paragraph 3. Thus there are six independent equations to determine 10 functions. Four of these functions may be freely chosen. (This freedom of choice is called *gauge freedom* and is an important feature of the physics of certain material interactions.)
5. So far we have assumed the coordinate fixing condition: $\sqrt{-g} = 1$. If $\sqrt{-g} \neq 1$, then the following invariant condition may be applied:

$$ds^2 = g_{\mu\nu} dx^\mu dx^\nu$$

or, in another form:

$$1 = g_{\mu\nu} \frac{dx^\mu}{ds} \frac{dx^\nu}{ds}$$

Also, in paragraph 3, the $S_{\mu\nu}$ of equation (30b) reappear in the general field equations.

6. If the kinematic motion is also required, then the geodesic equation (8) may be invoked.
7. Test the result empirically and/or its behavior under special conditions for which there are available reliable solutions. These include Newtonian approximations, STR conditions, and approximations to STR (post-Newtonian) conditions.
8. Simulations may be used to filter and tune candidate relativistic relationships.

R. Generally Covariant Formulation of Classical Electromagnetism

R1. Generally Covariant Formulation of Maxwell's Equations

Let Φ_ν be covariant components of the electromagnetic potential 4-vector. The curl of Φ_ν defines the covariant electromagnetic field tensor. Therefore applying equation (18) we get the following expression of this anti-symmetrical tensor:

$$F_{\rho\sigma} \overset{\text{def}}{=} \frac{\partial\Phi_\rho}{\partial x^\sigma} - \frac{\partial\Phi_\sigma}{\partial x^\rho}$$

Recalling equation (20), we may write:

$$F_{\rho\sigma\tau} = \frac{\partial F_{\rho\sigma}}{\partial x^\tau} + \frac{\partial F_{\sigma\tau}}{\partial x^\rho} + \frac{\partial F_{\tau\rho}}{\partial x^\sigma} = 0 \ : \rho, \sigma, \tau = 1, 2, 3, 4 \quad (48a)$$

The $F_{\rho\sigma\tau}$ vanish as the anti-symmetry of $F_{\rho\sigma}$ in ρ and σ and the commutativity of the partial derivative produce absolute mutual cancellation in the summation of $F_{\rho\sigma\tau}$ expressed in terms of coordinate derivatives of Φ_ν.

We can write out (48a) for the following substitutions of (ρ, σ, τ) as:

(ρ, σ, τ)

$$\left.\begin{array}{ll}
(1,2,3) & \dfrac{\partial F_{12}}{\partial x^3} + \dfrac{\partial F_{23}}{\partial x^1} + \dfrac{\partial F_{31}}{\partial x^2} = 0 \\[2ex]
(2,3,4) & \dfrac{\partial F_{23}}{\partial x^4} + \dfrac{\partial F_{34}}{\partial x^2} + \dfrac{\partial F_{42}}{\partial x^3} = 0 \\[2ex]
(3,4,1) & \dfrac{\partial F_{34}}{\partial x^1} + \dfrac{\partial F_{41}}{\partial x^3} + \dfrac{\partial F_{13}}{\partial x^4} = 0 \\[2ex]
(4,1,2) & \dfrac{\partial F_{41}}{\partial x^2} + \dfrac{\partial F_{12}}{\partial x^4} + \dfrac{\partial F_{24}}{\partial x^1} = 0
\end{array}\right\} (48b)$$

Identify:

$$\left.\begin{array}{lllll}
F_{23} = B_x & F_{31} = B_y & F_{12} = B_z & x^1 = x & x^3 = z \\
F_{14} = E_x & F_{24} = E_y & F_{34} = E_z & x^2 = y & x^4 = t
\end{array}\right\} (49)$$

where $\bar{\mathbf{B}}$ and $\bar{\mathbf{E}}$ are the magnetic and the electric 3-vector fields, respectively. Substituting (49) in equations (48b) and rearranging the array yield:

$$\frac{\partial B_x}{\partial x} + \frac{\partial B_y}{\partial y} + \frac{\partial B_z}{\partial z} = 0$$

$$-\frac{\partial B_x}{\partial t} = \frac{\partial E_z}{\partial y} - \frac{\partial E_y}{\partial z}$$

$$-\frac{\partial B_y}{\partial t} = \frac{\partial E_x}{\partial z} - \frac{\partial E_z}{\partial x}$$

$$-\frac{\partial B_z}{\partial t} = \frac{\partial E_y}{\partial x} - \frac{\partial E_x}{\partial y}$$

Thus written, the first equation recalls:

$$\nabla.\bar{\mathbf{B}} = 0$$

and the last three recall: $-\partial\bar{\mathbf{B}}/\partial t = \nabla \wedge \bar{\mathbf{E}}$

These are Maxwell's equations for charge-free regions of electromagnetic fields.

In the case of the presence of electromagnetic sources - charges and currents - introduce the anti-symmetric contravariant electromagnetic field tensor by raising the indices of the anti-symmetric covariant electromagnetic field tensor:

$$F^{\mu\nu} = g^{\mu\alpha} g^{\nu\beta} F_{\alpha\beta}$$

Recalling the divergence of the contravariant 6-vector given in equation (25), for $\sqrt{-g} = 1$, we may write:

$$J^{\mu} = \frac{\partial F^{\mu\nu}}{\partial x^{\nu}} \qquad \mu, \nu = 1, 2, 3, 4 \qquad (50a)$$

The J^{μ} are the sources of electromagnetic fluxes. Writing out (50a):

$$\left.\begin{aligned}
\frac{\partial F^{12}}{\partial x^2} + \frac{\partial F^{13}}{\partial x^3} + \frac{\partial F^{14}}{\partial x^4} &= J^1 \\
\frac{\partial F^{21}}{\partial x^1} + \frac{\partial F^{23}}{\partial x^3} + \frac{\partial F^{24}}{\partial x^4} &= J^2 \\
\frac{\partial F^{31}}{\partial x^1} + \frac{\partial F^{32}}{\partial x^2} + \frac{\partial F^{34}}{\partial x^4} &= J^3 \\
\frac{\partial F^{41}}{\partial x^1} + \frac{\partial F^{42}}{\partial x^2} + \frac{\partial F^{43}}{\partial x^3} &= J^4
\end{aligned}\right\} (50b)$$

Identify:

$$\left.\begin{array}{llllll}
F^{23} = H_x & F^{14} = -D_x & J^1 = j_x & J^4 = \rho & x^1 = x & x^4 = t \\
F^{31} = H_y & F^{24} = -D_y & J^2 = j_y & & x^2 = y & \\
F^{12} = H_z & F^{34} = -D_z & J^3 = j_z & & x^3 = z &
\end{array}\right\} (50c)$$

Where the 3-vectors $\bar{\mathbf{H}}$ and $\bar{\mathbf{D}}$ are the magnetizing field and the electrical displacement field, respectively. The J^{μ} - for $\mu = 1, 2, 3$ - are the components of the current density 3-vector $\bar{\mathbf{J}}$ and $J^4 = \rho$ is the charge density.

Substituting equations (50c) in (50b) yields:

$$\frac{\partial H_z}{\partial y} - \frac{\partial H_y}{\partial z} - \frac{\partial D_x}{\partial t} = j_x$$

$$-\frac{\partial H_z}{\partial x} + \frac{\partial H_x}{\partial z} - \frac{\partial D_y}{\partial t} = j_y$$

$$\frac{\partial H_y}{\partial x} - \frac{\partial H_x}{\partial y} - \frac{\partial D_z}{\partial t} = j_z$$

$$\frac{\partial D_x}{\partial x} + \frac{\partial D_y}{\partial y} + \frac{\partial D_z}{\partial z} = \rho$$

The first three equations recall:

$$\nabla \wedge \bar{\mathbf{H}} = \bar{\mathbf{J}} + \partial \bar{\mathbf{D}} / \partial t$$

and the last recalls: $$\nabla . \bar{\mathbf{D}} = \rho$$

These are Maxwell's equations for the general case with free charges.

Therefore, the equations:

$$\frac{\partial F_{\rho\sigma}}{\partial x^{\tau}} + \frac{\partial F_{\sigma\tau}}{\partial x^{\rho}} + \frac{\partial F_{\tau\rho}}{\partial x^{\sigma}} = 0$$

$$F^{\mu\nu} = g^{\mu\alpha} g^{\nu\beta} F_{\alpha\beta}$$

and $$\partial F^{\mu\nu} / \partial x_{\nu} = J^{\mu} \qquad \sqrt{-g} = 1$$

are the equivalent of Maxwell's equations and, as they are expressed in tensors, they are generally covariant.

R2. Formation of the Electromagnetic Energy Tensor

Consider the inner product:

$$\kappa_{\sigma} = F_{\sigma\mu} J^{\mu} \tag{51}$$

Writing out this equation yields:

$$\kappa_1 = F_{12}J^2 + F_{13}J^3 + F_{14}J^4$$
$$\kappa_2 = F_{21}J^1 + F_{23}J^3 + F_{24}J^4$$
$$\kappa_3 = F_{31}J^1 + F_{32}J^2 + F_{34}J^4$$
$$\kappa_4 = F_{41}J^1 + F_{42}J^2 + F_{43}J^3$$

Making the identifications of equations (49) and (50c), we get:

$$\left.\begin{aligned} \kappa_1 &= B_z j_y - B_y j_z + \rho E_x \\ \kappa_2 &= - B_z j_x + B_x j_z + \rho E_y \\ \kappa_3 &= B_y j_x - B_x j_y + \rho E_z \\ \kappa_4 &= -E_x j_x - E_y j_y - E_z j_z \end{aligned}\right\} \tag{52}$$

The first three equations yield:

$$(\kappa_1 \ \ \kappa_2 \ \ \kappa_3)^T = \bar{\mathbf{J}} \wedge \bar{\mathbf{B}} + \rho \bar{\mathbf{E}}$$

Here the superscript T indicates the transpose.

The expression on the right of the last equation is the Lorentz force density – the force exerted *on* the charges per unit 3-D volume by the electromagnetic field. Or, rephrased, this expression is the negative of the force exerted *by* the charges on the electromagnetic field per unit volume. Recalling that force is equivalent to the transitional rate of change of momentum, then the expression is equivalent to the negative momentum transferred from the charges to the electromagnetic field per unit 4-D volume.

The last equation of (52) may be written as:

$$\kappa_4 = -\bar{\mathbf{E}}.\,\bar{\mathbf{J}} \tag{53}$$

The negative of κ_4 is the energy that is transferred by the charges to the field per unit 4-D volume.

Therefore, the κ_σ are the negatives of the momentum densities and the energy density transferred from the electric charges to the electromagnetic field per unit duration. If the charges are free, then the κ_σ will vanish.

Substituting equation (50a) into (51) and rewriting:

$$\kappa_\sigma = F_{\sigma\mu}J^\mu = F_{\sigma\mu}\frac{\partial F^{\mu\nu}}{\partial x^\nu} = \frac{\partial}{\partial x^\nu}(F_{\sigma\mu}F^{\mu\nu}) - F^{\mu\nu}\frac{\partial F_{\sigma\mu}}{\partial x^\nu} \tag{54}$$

Considering the last term and recalling equation (48a), we may write:

$$F^{\mu\nu}\frac{\partial F_{\mu\nu}}{\partial x^\sigma} = -F^{\mu\nu}\frac{\partial F_{\sigma\mu}}{\partial x^\nu} - F^{\mu\nu}\frac{\partial F_{\nu\sigma}}{\partial x^\mu} \tag{55}$$

Interchange of summation indices μ and ν in the last term and the anti-symmetry of $F_{\mu\sigma}$ and $F^{\nu\mu}$ reveal the equality of both terms on the right, so we may write:

$$F^{\mu\nu}\frac{\partial F_{\mu\nu}}{\partial x^\sigma} = -2F^{\mu\nu}\frac{\partial F_{\sigma\mu}}{\partial x^\nu}$$

Substituting this into equation (54):

$$\kappa_\sigma = \frac{\partial}{\partial x^\nu}(F_{\sigma\mu}F^{\mu\nu}) + \frac{F^{\mu\nu}}{2}\frac{\partial F_{\mu\nu}}{\partial x^\sigma}$$

Lowering the indices in the second term on the right yields:

$$\kappa_\sigma = \frac{\partial}{\partial x^\nu}(F_{\sigma\mu}F^{\mu\nu}) + \frac{1}{2}g^{\mu\alpha}g^{\nu\beta}F_{\alpha\beta}\frac{\partial F_{\mu\nu}}{\partial x^\sigma}$$

In the first term, anti-symmetry of $F^{\mu\nu}$ in μ and ν and, in the second term, symmetry of $g^{\mu\nu}$ in the same indices and the interchange of summation indices α and μ on the one hand and, on the other, of β and ν allow us to write:

$$\begin{aligned}\kappa_\sigma &= -\frac{\partial}{\partial x^\nu}(F_{\sigma\mu}F^{\nu\mu}) + \frac{1}{4}g^{\mu\alpha}g^{\nu\beta}\left(F_{\alpha\beta}\frac{\partial F_{\mu\nu}}{\partial x^\sigma} + F_{\mu\nu}\frac{\partial F_{\alpha\beta}}{\partial x^\sigma}\right)\\ &= -\frac{\partial}{\partial x^\nu}(F_{\sigma\mu}F^{\nu\mu}) + \frac{1}{4}g^{\mu\alpha}g^{\nu\beta}\frac{\partial}{\partial x^\sigma}(F_{\alpha\beta}F_{\mu\nu})\end{aligned}$$

Rewrite as:

$$\kappa_\sigma = -\frac{\partial}{\partial x^\nu}(F_{\sigma\mu}F^{\nu\mu}) + \frac{1}{4}\left[\frac{\partial}{\partial x^\sigma}(g^{\mu\alpha}g^{\nu\beta}F_{\alpha\beta}F_{\mu\nu}) - F_{\alpha\beta}F_{\mu\nu}\frac{\partial}{\partial x^\sigma}(g^{\mu\alpha}g^{\nu\beta})\right]$$

In the square brackets, raising the indices in the first term and differentiating the second term yield:

$$\kappa_\sigma = -\frac{\partial}{\partial x^\nu}(F_{\sigma\mu}F^{\nu\mu}) + \frac{1}{4}\frac{\partial}{\partial x^\sigma}(F_{\alpha\beta}F^{\alpha\beta}) - \frac{1}{4}F_{\alpha\beta}F_{\mu\nu}\left(g^{\mu\alpha}\frac{\partial g^{\nu\beta}}{\partial x^\sigma} + g^{\nu\beta}\frac{\partial g^{\mu\alpha}}{\partial x^\sigma}\right)$$

Recalling (13) we may rewrite this as:

$$\kappa_\sigma = -\frac{\partial}{\partial x^\nu}(F_{\sigma\mu}F^{\nu\mu}) + \frac{1}{4}\frac{\partial}{\partial x^\sigma}(F_{\alpha\beta}F^{\alpha\beta}) + \frac{1}{4}F_{\alpha\beta}F_{\mu\nu}g^{\mu\alpha}g^{\nu\rho}g^{\beta\tau}\frac{\partial g_{\rho\tau}}{\partial x^\sigma}$$
$$+ \frac{1}{4}F_{\alpha\beta}F_{\mu\nu}g^{\nu\beta}g^{\mu\rho}g^{\alpha\tau}\frac{\partial g_{\rho\tau}}{\partial x^\sigma}$$

Raising indices in the third and fourth terms of the right hand side yields:

$$\kappa_\sigma = -\frac{\partial}{\partial x^\nu}(F_{\sigma\mu}F^{\nu\mu}) + \frac{1}{4}\frac{\partial}{\partial x^\sigma}(F_{\alpha\beta}F^{\alpha\beta}) + \frac{1}{4}F^{\mu\tau}F_{\mu\nu}g^{\nu\rho}\frac{\partial g_{\rho\tau}}{\partial x^\sigma}$$
$$+ \frac{1}{4}F^{\tau\nu}F_{\mu\nu}g^{\mu\rho}\frac{\partial g_{\rho\tau}}{\partial x^\sigma}$$

With the anti-symmetry of the field tensors, reversing the order of their indices in the third term and interchanging the summation indices μ and ν reveals the equality of the third and fourth terms, therefore we may write:

$$\kappa_\sigma = -\frac{\partial}{\partial x^\nu}(F_{\sigma\mu}F^{\nu\mu}) + \frac{1}{4}\frac{\partial}{\partial x^\sigma}(F_{\alpha\beta}F^{\alpha\beta}) + \frac{1}{2}F_{\nu\mu}F^{\tau\mu}g^{\nu\rho}\frac{\partial g_{\rho\tau}}{\partial x^\sigma}$$

Recalling (11b) we may rewrite the last term:

$$\kappa_\sigma = -\frac{\partial}{\partial x^\nu}(F_{\sigma\mu}F^{\nu\mu}) + \frac{1}{4}\frac{\partial}{\partial x^\sigma}(F_{\alpha\beta}F^{\alpha\beta}) - \frac{1}{2}F_{\nu\mu}F^{\tau\mu}g_{\rho\tau}\frac{\partial g^{\nu\rho}}{\partial x^\sigma}$$

Changing summation indices μ to α then ρ to μ yields:

$$\kappa_\sigma = -\frac{\partial}{\partial x^\nu}(F_{\sigma\alpha}F^{\nu\alpha}) + \frac{1}{4}\frac{\partial}{\partial x^\sigma}(F_{\alpha\beta}F^{\alpha\beta}) - \frac{1}{2}F_{\nu\alpha}F^{\tau\alpha}g_{\mu\tau}\frac{\partial g^{\nu\mu}}{\partial x^\sigma}$$

Recalling (36a) and (17) and for $\sqrt{-g} = 1$, this equation may be rewritten as:

$$\kappa_\sigma = \frac{\partial}{\partial x^\nu}\left(-F_{\sigma\alpha}F^{\nu\alpha} + \frac{1}{4}\delta^\nu_\sigma F^{\alpha\beta}F_{\alpha\beta}\right) + \frac{1}{2}g_{\mu\tau}\frac{\partial g^{\mu\nu}}{\partial x^\sigma}\left(-F_{\nu\alpha}F^{\tau\alpha} + \frac{1}{4}\delta^\tau_\nu F^{\alpha\beta}F_{\alpha\beta}\right)$$

Now letting:
$$T^\nu_\sigma = -F_{\sigma\alpha}F^{\nu\alpha} + \frac{1}{4}\delta^\nu_\sigma F^{\alpha\beta}F_{\alpha\beta} \qquad (56)$$

we may write:
$$\kappa_\sigma = \frac{\partial T^\nu_\sigma}{\partial x^\nu} + \frac{1}{2}g_{\mu\tau}\frac{\partial g^{\mu\nu}}{\partial x^\sigma}T^\tau_\nu \qquad :\sqrt{g} = -1$$

Lowering the index in the last term:

$$\kappa_\sigma = \frac{\partial T^{\nu}_{\sigma}}{\partial x^{\nu}} + \frac{1}{2}\frac{\partial g^{\mu\nu}}{\partial x^{\sigma}} T_{\mu\nu} \tag{57}$$

Recalling equation (28a) on page 35, if κ_σ vanishes, as it does for charges free of influences other than the electromagnetic and gravitational fields, then κ_σ is the divergence of the *electromagnetic energy tensor* T^{ν}_{σ}.

Recalling the discussion in § P, the T^{4}_{σ} - for $\sigma = 1, 2, 3$ - represent the electromagnetic momentum densities collinear with x^1, x^2, and x^3, respectively. Now $\delta^{4}_{\sigma} = 0$ for $\sigma \neq 4$ and the field tensors are anti-symmetric, so equation (56) may be partially written out as:

$$T^4_1 = -F_{12}F^{42} - F_{13}F^{43} \qquad T^4_2 = -F_{21}F^{41} - F_{23}F^{43} \qquad T^4_3 = -F_{31}F^{41} - F_{32}F^{42}$$

With the identifications of (49) and (50c) and as $\bar{\mathbf{D}} = \varepsilon_o\bar{\mathbf{E}}$ and $\bar{\mathbf{B}} = \mu_o\bar{\mathbf{H}}$ where ε_σ and μ_o are the permittivity and permeability of the vacuum, respectively, we get:

$$\left.\begin{aligned} T^4_1 &= -B_zD_y + B_yD_z = (-H_zE_y + H_yE_z)\mu_o\varepsilon_\sigma \\ T^4_2 &= B_zD_x - B_xD_z = (H_zE_x - H_xE_z)\mu_o\varepsilon_\sigma \\ T^4_3 &= (-B_yD_x + B_xD_y) = (-H_yE_x + H_xE_y)\,\mu_o\varepsilon_\sigma \end{aligned}\right\} \tag{58}$$

So the Maxwell-Poynting vector (in mass density-velocity units) is:

$$\bar{\mathbf{S}} = \bar{\mathbf{E}} \wedge \bar{\mathbf{H}} = \begin{vmatrix} \hat{\mathbf{\imath}} & \hat{\mathbf{\jmath}} & \hat{\mathbf{k}} \\ E_x & E_y & E_z \\ H_x & H_y & H_z \end{vmatrix} = \hat{\mathbf{\imath}}\left(E_yH_z - E_zH_y\right) + \hat{\mathbf{\jmath}}(E_zH_x - E_xH_z) + \hat{\mathbf{k}}(E_xH_y - E_yH_x)$$

$$= -\frac{(\hat{\mathbf{\imath}}T^4_1 + \hat{\mathbf{\jmath}}T^4_2 + \hat{\mathbf{k}}T^4_3)}{\mu_\sigma\varepsilon_\sigma} = -c^2(\hat{\mathbf{\imath}}T^4_1 + \hat{\mathbf{\jmath}}T^4_2 + \hat{\mathbf{k}}T^4_3)$$

as the speed of light, in a vacuum, is given by $c = 1/(\mu_\sigma\varepsilon_\sigma)^{\frac{1}{2}}$. So we can write:

$$-(\hat{\mathbf{\imath}}T^4_1 + \hat{\mathbf{\jmath}}T^4_2 + \hat{\mathbf{k}}T^4_3) = \frac{\bar{\mathbf{S}}}{c^2} \tag{59}$$

Therefore, the T^4_σ for $\sigma = 1, 2, 3$ are (in energy density-velocity units) the negatives of the *electromagnetic field momentum densities*[21] of the Maxwell-Poynting vector. The T^4_σ for $\sigma = 1, 2, 3$ are the relativistic momentum densities imparted *by* the charges to the field. Conservation of momentum implies that the charges' loss of momentum is the field's gain. (Based on the discussion in § O, the conservation of momentum holds only in a negligible gravitational field. This condition is implicit in Maxwell's theory of electromagnetism.)

Similar substitutions in the expression of the relativistic energy density tensor yield:

$$T_4^4 = -F_{4\alpha}F^{4\alpha} + \frac{1}{4}F_{\alpha\beta}F^{\alpha\beta} = (\epsilon_o \bar{\mathbf{E}}^2 + \mu_o \bar{\mathbf{H}}^2)/2 \qquad (60)$$

The term on the right is the expression, given in Poynting's theorem, of the electromagnetic energy density.

S. Gravity, Bodies and Light

S1. Energy Tensors of Fluids

For agglomerations of particles with only gravitational interaction (sometimes referred to as *dust*), the contravariant energy tensor is:

$$T^{\alpha\beta} = \rho u^\alpha u^\beta \qquad (61a)$$

where the u^α are components of a 4-vector velocity - dx^α/ds - of the fluid as measured in the observer's chart. The quantity ρ is the fluid's 3D energy density as measured in the fluid's rest frame and ds is the proper infinitesimal transitional interval measured in the observer's frame.

Consider the energy tensor T^{11}:

$$T^{11} = \rho\left(\frac{dx^1}{ds}\right)^2 = \rho(u^1)^2$$

Multiply the numerator and denominator of the right side by dx^2dx^3 and rewrite:

$$T^{11} = [\,(\rho dx^1dx^2dx^3u^1)/ds]/dx^2dx^3$$

The expression in parenthesis is the momentum along x^1 of the energy flowing through the infinitesimal 3-volume $dx^1dx^2dx^3$ during the transitional interval ds. If the movement is brought to rest during ds, then a force (expressed in the square brackets) is experienced due to the transitional rate of change of momentum. So T^{11} is the 'pressure' exerted along x^1 through the infinitesimal surface element dx^2dx^3 [cf. § P]. And similar relationships attend T^{22} and T^{33}.

Now consider the energy tensor T^{14}:

$$T^{14} = \rho\frac{dx^1}{ds}\frac{dx^4}{ds}$$

The product of the first two factors on the right side is the momentum density of the dust translating along x^1. The last factor is a dimensionless ratio of the proper and coordinate transitional intervals.* So T^{14} is a component of the momentum.

* In STR conditions, a proper length dl, due to motion, is measured as being $dx^1 = \gamma dl$ long and a proper duration of ds is observed to be $dx^4 = ds/\gamma$. Then the product $dx^1dx^4 = dl\,ds$. With this substitution, T^{14} shows itself, in this case, to be the momentum density measurable in proper quantities: $\rho\,dl/ds$. So here dx^4/ds may be regarded as a 'scale factor'.

Now consider the case when the observer is at rest relative to the dust, that is, $u^1 = u^2 = u^3 = 0$ and $u^4 = 1$. This last equation results from substituting the previous equations into (1c). So, with these determinations, equation (61a) yields $T^{44} = \rho$.

For fluids in thermodynamic equilibrium – adiabatic and frictionless - referred to as *perfect fluids,* the energy tensor may be expressed as:

$$T^{\alpha\beta} = (\rho + p)u^\alpha u^\beta - pg^{\alpha\beta} \quad (61b)$$

Here p is the pressure and is expressed in the units of 3D energy density. In the fluid's rest frame, $u^\alpha = (0,\ 0,\ 0, 1)$ so $u^\alpha u^\beta = 1$ for $\alpha = \beta = 4$ and vanishes otherwise. Furthermore, the reciprocal tensors of the metric of equation (1c) are:

$$g^{\alpha\beta} = -1 \qquad :\text{for } \alpha = \beta = 1, 2, \text{ and } 3$$
$$g^{44} = 1$$
$$g^{\alpha\beta} = 0 \qquad :\text{for } \alpha \neq \beta.$$

So with these determinations equation (61b) yields:

$$T^{\alpha\beta} = p \qquad :\text{for } \alpha = \beta = 1, 2, \text{ and } 3$$
$$T^{44} = \rho$$
$$T^{\alpha\beta} = 0 \qquad :\text{for } \alpha \neq \beta$$

These results are as expected based upon the discussion in § P and assumed thermodynamic and STR conditions.

Given their tensorial characters, the energy tensor given in equations (61a) and (61b) remain valid in the more general cases where the metric tensor $g_{\alpha\beta}$ are functions of the coordinates. Dust and perfect fluid models are important for large scale modeling of cosmological and astrophysical processes.

The energy tensor of an imperfect fluid may be expressed in the form:

$$T^{\alpha\beta} = (\rho + p)u^\alpha u^\beta - pg^{\alpha\beta} + \pi^{\alpha\beta} + u^\alpha q^\beta + u^\beta q^\alpha \qquad u^\alpha q_\alpha = 0 \qquad u^\alpha \pi_{\alpha\beta} = 0$$

Here $\pi^{\alpha\beta}$ are the viscous shear stress components and the q^α are heat fluxes. As discussed in § P, the viscous shear stresses and the heat fluxes are transverse energy fluxes and this is made explicit by the last two equations.

S2. Weak Gravity of Slowly Moving Bodies: The Newtonian Approximation

Consider the case of a material point - a body of relatively infinitesimal mass and volume - moving at a low velocity and under the influence of a low-intensity and quasi-static gravitational field produced by a relatively massive gravitating body of mass-energy density ρ that moves at a low 3-D velocity. Assume the signature of the metric of the equation of the (almost) Minkowskian linear element to be $- - - +$.

A 3-vector velocity $dx^\mu/d\tau$ - $\mu = 1, 2, 3$ - of small magnitude implies:

$$(dx^\mu/d\tau)/c = dx^\mu/ds \cong 0 \qquad :\mu = 1, 2, 3$$

Here $ds = cd\tau$, $d\tau$ being the proper transitional interval in seconds as measured (locally) on the small body and c is the speed of light. This modifies the (almost) Minkowskian equation of the linear element - approximating equation (1c) - so that it may be rewritten as:

$$dx^4/ds \cong 1$$

The small body is unrestricted and as such its trajectory describes a geodesic. Under these conditions - that is, with substitutions from the two equations above - the geodesic equation (8) of the material point is modified to become:

$$\frac{d^2x^\tau}{dx^{4^2}} = \Gamma^\tau_{44} \qquad :\tau = 1, 2, 3 \qquad (62a)$$

Substituting from equation (7) yields:

$$\frac{d^2x^\tau}{dx^{4^2}} = -\frac{g^{\tau\alpha}}{2}\left(\frac{\partial g_{4\alpha}}{\partial x^4} + \frac{\partial g_{4\alpha}}{\partial x^4} - \frac{\partial g_{44}}{\partial x^\alpha}\right) \qquad :\alpha = 1, 2, 3, 4 \quad \tau = 1, 2, 3$$

Now, as approximately: $g_{\tau\sigma} = -1$ $\qquad :\tau = \sigma = 1, 2, 3$

and recalling that: $g_{\tau\sigma}g^{\tau\alpha} = 1$ $\qquad :\alpha = \sigma = 1, 2, 3$

$= 0$ $\qquad :\alpha \neq \sigma$

we get: $g^{\tau\alpha} = -1$ $\qquad :\alpha = \tau = 1, 2, 3$

Furthermore, the quasi-static nature of the gravitational field implies:

$$\frac{\partial g_{\mu\nu}}{\partial x^4} \cong 0 \qquad : \mu, \nu = 1, 2, 3, 4$$

So with $\alpha = \tau$, the geodesic equation becomes:

$$\frac{d^2x^\tau}{dx^{4^2}} = -\frac{1}{2}\frac{\partial g_{44}}{\partial x^\tau} \qquad :\tau = 1, 2, 3 \qquad (62b)$$

Set: $\qquad x^1 = x \qquad x^2 = y \qquad x^3 = z \qquad x^4 = ct$

so in Cartesian coordinates, the three equations of (62b) may be combined as:

$$\frac{d^2x}{dt^2}\hat{\mathbf{i}} + \frac{d^2y}{dt^2}\hat{\mathbf{j}} + \frac{d^2z}{dt^2}\hat{\mathbf{k}} = -c^2\left(\frac{1}{2}\frac{\partial g_{44}}{\partial x}\hat{\mathbf{i}} + \frac{1}{2}\frac{\partial g_{44}}{\partial y}\hat{\mathbf{j}} + \frac{1}{2}\frac{\partial g_{44}}{\partial z}\hat{\mathbf{k}}\right) \qquad (62c)$$

So written, the geodesic equation recalls Newton's law of gravity in the form:

$$\bar{\mathbf{a}}_g = -\nabla\Phi \qquad (62d)$$

with $\bar{\mathbf{a}}_g$ being the strength of the gravitational field (the relative gravitational acceleration of the bodies) and $\nabla\Phi$ being the gradient of the Newtonian gravitational potential Φ. Identity of equations (62c) and (62d) requires:

$$\nabla(c^2 g_{44}/2) = \nabla\Phi = \nabla(k+\Phi) \text{ for } \nabla k = 0$$

So:

$$g_{44} = 2(k + \Phi)/c^2$$

The quantity k is a constant of integration.

Recall the expression of Newton's gravitational potential (in units of energy per unit mass):

$$\Phi = -\frac{GM}{r} \tag{63a}$$

In the weak field, the potential is a small quantity. Therefore, the quantity $2\Phi/c^2$ - twice the potential in units of energy per unit energy - is extremely small. The field is almost Minkowskian with $g_{44} \cong 1$ so we require $2k/c^2 = 1$. This yields:

$$g_{44} = 1 + 2\Phi/c^2 \tag{63b}$$

Assuming the gravitating body has an energy tensor of the form given in equations (61a), consider the factors of induction of the gravitational field:

1. As for the small body, a low 3-vector velocity implies $u^\lambda \cong 0$ for $\lambda = 1, 2, 3$ and $u^4 \cong 1$. So equation (61a) yields $T^{\mu\nu} = T^{4\nu} = T^{\mu 4} = 0$ for $\mu, \nu = 1, 2, 3$ and $T^{44} = \rho$. Now lowering the indices of T^{44}:

$$T_{\mu\nu} = g_{\mu 4} g_{\nu 4} T^{44} = T_{44} = \rho$$

   as $g_{\lambda 4} \cong 0$ for $\lambda \neq 4$ and $g_{44} \cong 1$. Also, as $g_{44} \cong 1$, metric contraction of $T^{\mu\nu}$ yields $T = g_{\mu\nu} T^{\mu\nu} = g_{44} T^{44} = T^{44} = \rho$.
2. The low intensity of the induced field implies that terms in the field components - $\Gamma^\tau_{\mu\nu}$ - higher than those of the first order are insignificant. Thus the second term on the left of the induction equation (42), being a product of gravitational field components of the first order of magnitude, is negligible. This is the 'first approximation' referred to in sub-paragraph d of § K
3. The quasi-static nature of the field implies that the transitional derivatives of the components of the field almost vanish:

$$\frac{\partial \Gamma^4_{\mu\nu}}{\partial x^4} \cong 0 \qquad : \mu, \nu = 1, 2, 3, 4$$

4. Furthermore, according to equation (62a), in the conditions under consideration, the field components that are relevant are the Γ^τ_{44} for $\tau = 1, 2, 3$.

In light of these considerations, the relationship of gravitational induction given in equation (42) becomes:

$$\frac{\partial \Gamma^{\tau}_{44}}{\partial x^{\tau}} = -\kappa\left(T_{44} - \frac{g_{44}T}{2}\right) = -\frac{\kappa\rho}{2} \qquad :\sqrt{-g}=1,\ \tau=1,2,3$$

Writing out fully:

$$\frac{\partial \Gamma^{1}_{44}}{\partial x^{1}} + \frac{\partial \Gamma^{2}_{44}}{\partial x^{2}} + \frac{\partial \Gamma^{3}_{44}}{\partial x^{3}} = -\frac{\kappa\rho}{2} \qquad :\sqrt{-g}=1$$

In the equation above, substituting Γ^{τ}_{44} from equations (62a) and (62b) and Cartesian coordinates for the x^{μ} - for $\mu = 1, 2, 3$ - yield:

$$\frac{1}{2}\left(\frac{\partial^2 g_{44}}{\partial x^2} + \frac{\partial^2 g_{44}}{\partial y^2} + \frac{\partial^2 g_{44}}{\partial z^2}\right) = \frac{\kappa\rho}{2}$$

Substituting equation (63b) in the above equation yields:

$$\nabla^2\Phi = \kappa\rho c^2/2$$

The aspects of $\mathcal{R}_{\mu\nu}$, outlined in sub-paragraphs b and c of § K, expressed through (42) in the weak quasi-static field lead to this equation that with the identification

$$\kappa\rho c^2/2 = 4\pi G\gamma$$

recalls Poisson's form of Newton's law of gravity. Here G is the gravitational constant and γ is the mass density of the gravitating body.

From the foregoing, it is clear that Newtonian science of gravity occurs as a special case in Einstein's science of gravity.

Furthermore, by means of the last identification and the equation $\rho = \gamma c^2$, the constant in the general field equations is defined as:

$$\kappa = 8\pi G/c^4$$

So equation (42) becomes:

$$\frac{\partial \Gamma^{\alpha}_{\mu\nu}}{\partial x^{\alpha}} + \Gamma^{\alpha}_{\mu\beta}\Gamma^{\beta}_{\nu\alpha} = -\frac{8\pi G}{c^4}\left(T_{\mu\nu} - \frac{g_{\mu\nu}T}{2}\right) \qquad : \sqrt{-g} = 1$$

These sixteen equations are called the *Einstein Field Equations.*

S3. <u>Coordinate Intervals in a Weak Radial Gravitational Field</u>

Equation (63b) implies a gravitational influence on the determination of the transitional coordinate interval. For $ds^2 = 1$ and $dx^1 = dx^2 = dx^3 = 0$, the equation of the linear element approximates:

$$1 = g_{44}(dx^4)^2 = (1+2\Phi/c^2)(dx^4)^2$$

So to a first approximation[22] in $1/c^2$:

$$dx^4 = (1 + 2\Phi/c^2)^{-\frac{1}{2}} \cong (1 - \Phi/c^2)$$

As the gravitational potential is negative, the transitional coordinate interval becomes slightly greater in the presence of gravity.

With this change in g_{44} from the Minkowskian value of unity, in order to maintain $\sqrt{-g} = 1$ other terms of the metric are also altered. To a first approximation in $1/c^2$, such an alteration may be provided by a modified form of Einstein's radially symmetric equation of the linear element[23] with its metric specified as:

$$\left.\begin{aligned} g_{\rho\sigma} &= -\delta^{\rho}_{\sigma} + \frac{2\Phi x^{\rho} x^{\sigma}}{c^2 r^2} && : \rho, \sigma = 1, 2, 3 \\ g_{\rho 4} &= g_{4\rho} = 0 && : \rho = 1, 2, 3 \\ g_{44} &= 1 + 2\Phi/c^2 && \\ \text{where} \quad r &= \sqrt{(x^1)^2 + (x^2)^2 + (x^3)^2} && \end{aligned}\right\} \quad (64)$$

Consider a rod of unit measure $ds^2 = -1$ placed along the x^1 axis. This implies:

$$x^2 = x^3 = 0 \quad x^1 = r \quad dx^2 = dx^3 = dx^4 = 0$$

Therefore the equation of the linear element becomes:

$$-1 = -\left(1 - \frac{2\Phi}{c^2}\right)(dx^1)^2$$

and, to a first approximation in $1/c^2$, we get:

$$dx^1 \cong 1 + \Phi/c^2$$

The same holds for rods placed along the similarly radial x^2 and x^3 axes. As Φ is a negative quantity, then a radial coordinate interval is contracted.

For a rod of unit measure $ds^2 = -1$ laid perpendicular to the radial x^1 axis and parallel to the x^2 axis:

$$x^1 = r \quad x^2 = x^3 = 0 \quad dx^1 = dx^3 = dx^4 = 0.$$

Hence the equation of the linear element becomes:

$$-1 = -(dx^2)^2$$

So:

$$dx^2 = 1$$

Again, the same holds for rods placed transverse to the similarly radial x^2 and x^3 axes. Hence tangential coordinate intervals are unaffected by the field.

S4. The Schwarzschild Solution

Transforming equation (1c) to spherically symmetric coordinates and scaling the transitional intervals in seconds by means of the following substitutions:

$x^1 = r\sin\theta\cos\phi \qquad x^2 = r\sin\theta\sin\phi \qquad x^3 = r\cos\theta \qquad x^4 = ct \qquad s = c\tau$

We get:

$$-(cd\tau)^2 = dr^2 + r^2d\theta^2 + r^2\sin^2\theta d\phi^2 - (cdt)^2$$

This is an equation of the Minkowskian linear element where r, θ, and ϕ are the radial, colatitudinal (complement of the latitudinal), and longitudinal coordinates, respectively. The quantity τ is the proper duration in seconds and c is the speed of light.

The interval dr is radial while the mutually orthogonal arcual intervals - $rd\theta$ and $r\sin\theta d\varphi$ - are tangential. In light of the preceding considerations, in a weak radial gravitational field, the equation of the linear element becomes:

$$-(cd\tau)^2 = (1 - 2\Phi/c^2)dr^2 + r^2d\theta^2 + r^2\sin^2\theta d\phi^2 - (1 + 2\Phi/c^2)(cdt)^2 \qquad (65a)$$

This equation preserves, proportionately, the contraction of the radial coordinate interval and the dilation of the transitional interval previously arrived at. The tangential intervals, as before, are not influenced by the field. In the weak field, components of the metric with differing indices are here regarded as negligible.

Now to a first approximation in $1/c^2$:

$$(1-2\Phi/c^2) \cong (1+ 2\Phi/c^2)^{-1}$$

With this approximation and equation (63a) we may write:

$$-(cd\tau)^2 = \left(1 - \frac{2GM}{rc^2}\right)^{-1}(dr)^2 + r^2(d\theta)^2 + r^2\sin^2\theta(d\phi)^2 - \left(1 - \frac{2GM}{rc^2}\right)(cdt)^2$$

Or:

$$-(cd\tau)^2 = \left(1 - \frac{r_s}{r}\right)^{-1}(dr)^2 + r^2(d\theta)^2 + r^2\sin^2\theta(d\phi)^2 - \left(1 - \frac{r_s}{r}\right)(cdt)^2 \qquad (65b)$$

where the quantity $r_s = 2GM/c^2$ is called the *Schwarzschild radius.*

Equation (65b), called the *Schwarzschild solution,* defines a linear element in a radial gravitational field at distances greater than r_s from the centre of the gravitating body. It is a solution of the field equation of gravity as it specifies the required metric that describes the field components - $\Gamma^{\alpha}_{\mu\nu}$ - of equation (42).

S5. Light in Weak Gravity

S5.1 Deflection of Light from Around a Gravitating Body

Recall that Huyghens' Principle in its application, for example, to the refraction of light at an interface between two media yields the relationship:

$$\frac{\sin\phi_2}{\sin\phi_1} = \frac{v_2}{v_1} \tag{66a}$$

where the indexes 1 and 2 denote the different media, the angles ϕ_j are taken with respect to the normal to the interface, and the quantities v_j are the speeds of light in the different media. This relationship underpins Snell's law.

Consider this relationship in the context of an infinitesimal change of velocity that leads to a similarly infinitesimal change of direction. That is, let:

$$v_1 = v \qquad v_2 = v + dv$$

and

$$\phi_1 = \phi \qquad \phi_2 = \phi + d\phi$$

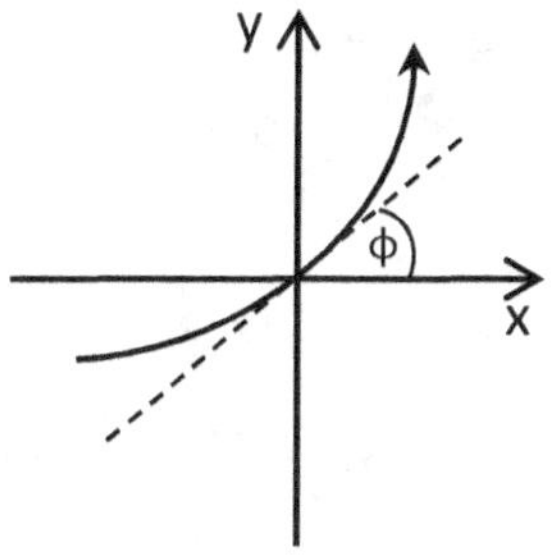

Fig. 1. Light Deflection

So that:
$$\frac{\sin(\phi + d\phi)}{\sin\phi} = \frac{v + dv}{v}$$

Or:
$$\frac{\sin\phi\,\cos(d\phi) + \sin(d\phi)\,\cos\phi}{\sin\phi} = \frac{v + dv}{v}$$

However, $\cos(d\phi) \cong 1$ and $\sin(d\phi) \cong d\phi$, so the equation above becomes:

$$1 + \frac{d\phi\,\cos\phi}{\sin\phi} = 1 + \frac{dv}{v}$$

This may be rewritten as:

$$d\phi\,\cot\phi = dv/v$$

This relationship usefully may be illustrated by Figure 1. The bending occurs in the (x, y) plane and $\cot\phi = dx/dy$. So we may rewrite the above equation as:

$$d\phi\frac{dx}{dy} = \frac{dv}{v} \qquad \text{or} \qquad \frac{d\phi}{dy} = \frac{1}{v}\frac{\partial v}{\partial x}$$

The partial derivation is incurred due to the velocity being a function of both the x and y coordinates.

Now $v = dr/dt \cong c$ where c is the speed of light in a vacuum and, importantly here as we shall see, in the absence of gravity. We may make the following substitutions $dx = dx^1$, $dy = dx^2$ and $dx^4 = cdt$ and rewrite the above equation as:

$$\frac{d\phi}{dx^2} = \frac{\partial\gamma}{\partial x^1} \tag{66b}$$

where:

$$\gamma = \frac{v}{c} = \frac{dr}{dx^4} = \left[\left(\frac{dx^1}{dx^4}\right)^2 + \left(\frac{dx^2}{dx^4}\right)^2\right]^{1/2} \tag{67}$$

is the speed of light scaled in units such that $c = 1$, and $dr = \sqrt{(dx^1)^2 + (dx^2)^2}$.

Equations (66a) and (66b) are finite and infinitesimal expressions, respectively, of Huyghen's principle corresponding to discontinuous and continuous changes, respectively, of the speed of light due to corresponding changes in the conditions of propagation.

From STR we know that setting $ds = 0$ in the equation of Minkowski's linear element - for example, as given in equation (1c) - yields the speed of light. A 4-D trajectory satisfying this condition is called a *null geodesic*. So we may write:

$$ds^2 = -(dx^1)^2 - (dx^2)^2 - (dx^3)^2 + (dx^4)^2 = 0$$

rewriting:

$$\gamma^2 = \left(\frac{dr}{dx^4}\right)^2 = \frac{(dx^1)^2 + (dx^2)^2 + (dx^3)^2}{(dx^4)^2} = 1$$

which with equation (67) yields $v = c$.

Therefore equation (3) implies that the speed of light, in general, is implicit in:

$$g_{\mu\nu}dx^\nu dx^\mu = 0 \qquad \mu, \nu = 1, 2, 3, 4$$

In the given frame of reference, $x^3 = 0$. So, ignoring the very small $g_{\mu\nu}$ for $\mu \neq \nu$:

$$g_{11}(dx^1)^2 + g_{22}(dx^2)^2 + g_{44}(dx^4)^2 = 0$$

Rewrite this as:

$$\left(\frac{dx^2}{dx^4}\right)^2 = -\frac{1}{g_{22}}\left[g_{11}\left(\frac{dx^1}{dx^4}\right)^2 + g_{44}\right] \tag{68}$$

and substitute into equation (67):

$$\gamma = \sqrt{\left(\frac{dx^1}{dx^4}\right)^2 - \frac{1}{g_{22}}\left[g_{11}\left(\frac{dx^1}{dx^4}\right)^2 + g_{44}\right]} = \left[\left(1 - \frac{g_{11}}{g_{22}}\right)\left(\frac{dx^1}{dx^4}\right)^2 - \frac{g_{44}}{g_{22}}\right]^{\frac{1}{2}} \tag{69}$$

Now recalling the metric tensors given in equations (64), we may write:

$$1 - \frac{g_{11}}{g_{22}} = 1 - \frac{-1 + \frac{2\Phi(x^1)^2}{c^2 r^2}}{-1 + \frac{2\Phi(x^2)^2}{c^2 r^2}} \cong 0$$

So squaring (69) and substituting the above approximation:

$$\gamma^2 = -\frac{g_{44}}{g_{22}} = \frac{1 + \frac{2\Phi}{c^2}}{1 - \frac{2\Phi(x^2)^2}{c^2 r^2}} = 1 + \frac{\frac{2\Phi}{c^2} + \frac{2\Phi(x^2)^2}{c^2 r^2}}{1 - \frac{2\Phi(x^2)^2}{c^2 r^2}} \cong 1 + \frac{2\Phi}{c^2}\left[1 + \frac{(x^2)^2}{r^2}\right]$$

to a first approximation in $1/c^2$. Take the square root to the same approximation and substitute $\Phi = -GM/r$:

$$\gamma = 1 - \frac{GM}{c^2}\left[\frac{1}{r} + \frac{(x^2)^2}{r^3}\right] \tag{70}$$

This implies that the speed of light, in a gravitational field, is always less than c.

Substituting the above equation into (66b) and differentiating yield:

$$\frac{d\phi}{dx^2} = \frac{\partial \gamma}{\partial x^1} = -\frac{GM}{c^2}\left[\frac{\partial}{\partial x^1}\left(\frac{1}{r}\right) + (x^2)^2 \frac{\partial}{\partial x^1}\left(\frac{1}{r^3}\right)\right] = \frac{GM}{c^2}\left[\frac{x^1}{r^3} + (x^2)^2 \frac{3x^1}{r^5}\right]$$

Consider a beam of light passing the sun at a minimum distance of b. This is depicted in Figure 2. The presence of the massive body fundamentally changes the nature of the coordinate system, even though we continue to denote the coordinates as before. Now multiply the above equation by dx^2 and integrate as the ray goes from $x^2 = -\infty$ to $x^2 = \infty$ along $x^1 = b$:

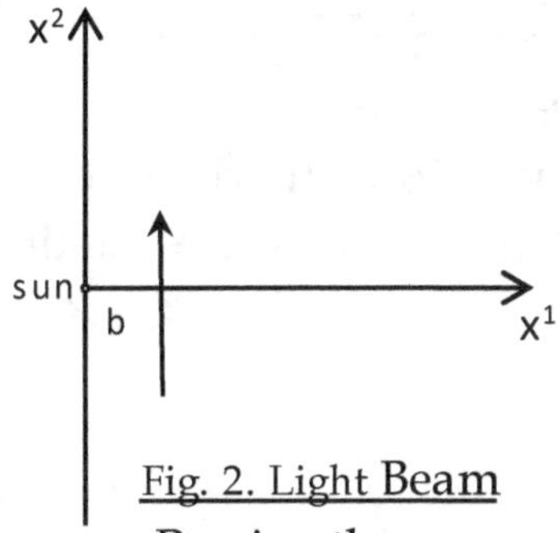

Fig. 2. Light Beam Passing the sun

$$\Delta\phi = \frac{GMb}{c^2} \int_{-\infty}^{\infty} \left[\frac{1}{r^3} + \frac{3(x^2)^2}{r^5}\right] dx^2$$

$$= \frac{GM}{bc^2}\left[\frac{x^2}{r} + \frac{(x^2)^3}{r^3}\right]\Bigg|_{x^2=-\infty}^{x^2=\infty} = \frac{4GM}{bc^2}$$

The positive/negative deflection corresponding to a positive/negative value of b implies that the ray becomes more concave with respect to the sun. The gravitational deflection of light is a prediction of Einstein published in his *FGTR* and has been verified to a high degree of precision.

Fundamentally, Huyghen's principle relates a variation of the beam's speed across its propagating wavefront to a resulting change of direction of propagation. Such a variation familiarly occurs in refraction resulting from a discontinuous change in the refractive index of the optical media that, relatively, traverses the geometrical surface of a propagating wavefront due to an oblique incidence of the beam into the optical interface. However, in the case under consideration, the optical medium is the isotropic vacuum with unit refractive index everywhere. Here, as the beam passes the sun, it experiences a gravitationally induced speed reduction, the effect of which declines across its wavefront as the distance from the sun increases - as given in equation (70). This gives rise to a smooth rotation of the propagating wavefront and, thereby, to a smooth deflection of its 'outer-pointing' normal - the direction of which is the ray's – in such a sense that there results the concavity of the locus of the beam with respect to the sun.

Since the vacuum is not optically dispersive, then the deflection is not frequency dependent and no spectrum can result as may happen in refractions.

Note that this deflection occurs despite the fact that the beam travels along a locus described by $x^1 = b$. This reveals the curvilinear nature of the coordinates in the radial gravitational field. The expression of x^2 as a function of x^1 - as may be shown through equation (68) - reveals that, here, x^2 is not orthogonal to x^1.

S5.2 Delay of Light

Equation (70) implies an increased duration of travel for light in the gravitational field compared to its travel in a gravity-free vacuum.

Setting ds = 0 in equation (65a), consider the null geodesic given in:

$$(1 - 2\Phi/c^2)\,(dr)^2 + r^2(d\theta)^2 + r^2\sin^2\theta(d\phi)^2 - (1 + 2\Phi/c^2)(cdt)^2 = 0$$

Rewrite as:

$$\left(1 + \frac{2\Phi}{c^2}\right)(cdt)^2 = \left(1 - \frac{2\Phi}{c^2}\right)[\,(dr)^2 + r^2(d\theta)^2 + r^2\sin^2\theta(d\phi)^2]$$
$$+ \frac{2\Phi}{c^2}\left[r^2(d\theta)^2 + r^2\sin^2\theta(d\phi)^2\right]$$

As the magnitude of $2\Phi/c^2 \ll 1$, then the second term on the right is quite small compared to the first and may be ignored, so we may write:

$$\left(1 + \frac{2\Phi}{c^2}\right)(cdt)^2 = \left(1 - \frac{2\Phi}{c^2}\right)[\,(dr)^2 + r^2(d\theta)^2 + r^2\sin^2\theta(d\phi)^2] \qquad (71a)$$

Rewrite the equation above as:

$$\left(1+\frac{2\Phi}{c^2}\right)(cdt)^2 = \left(1-\frac{2\Phi}{c^2}\right)(d\sigma)^2$$

where

$$d\sigma = \sqrt{(dr)^2 + r^2(d\theta)^2 + r^2\sin^2\theta(d\phi)^2}$$

is the differential path length. Hence we may write:

$$(cdt)^2 = \frac{\left(1-\frac{2\Phi}{c^2}\right)}{\left(1+\frac{2\Phi}{c^2}\right)}(d\sigma)^2 \cong \left(1-\frac{2\Phi}{c^2}\right)^2 (d\sigma)^2$$

and taking the square root and substituting $\Phi = -GM/r$:

$$dt = \frac{1}{c}\left(1+\frac{2GM}{c^2 r}\right)d\sigma \qquad (71b)$$

The duration of travel may be obtained by integrating the above equation along the light's path.

Consider the case where the path is linear. This is depicted in Figure 3 where the light's path is along the line BC. The origin of the frame of Figure 3 is at the centre of the gravitating body. For simplification, let the coordinate system be fixed so that along the linear locus of the ray the coordinate ϕ does not vary. This implies $d\phi = 0$. So the infinitesimal path length may take the following form:

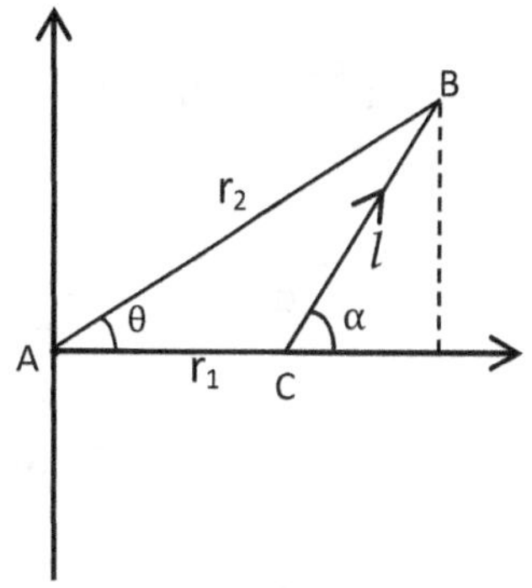

Fig. 3. Non-radial Light Path

$$d\sigma = \sqrt{(dr)^2 + r^2(d\theta)^2} = \left[r^2 + \left(\frac{dr}{d\theta}\right)^2\right]^{1/2} d\theta$$

So integrating equation (71b) over the light's path:

$$\Delta t = \frac{1}{c}\int_0^l d\sigma + \frac{2GM}{c^3}\int_0^\theta \frac{1}{r}\left[r^2 + \left(\frac{dr}{d\theta}\right)^2\right]^{1/2} d\theta \qquad (71c)$$

Represent (in Cartesian coordinates) the light path by the equation:

$$y = mx - k$$

where:

$$m = dy/dx = \tan\alpha$$

So by virtue of Pythagoras' theorem we may write:

$$\sin\alpha = \frac{m}{\sqrt{1+m^2}} \quad \text{and} \quad \cos\alpha = \frac{1}{\sqrt{1+m^2}} \tag{72}$$

Transforming the equation of the light's path to spherical coordinates by the substitutions:

$$y = r\sin\theta \quad \text{and} \quad x = r\cos\theta$$

and we get:

$$r = -k/(\sin\theta - m\cos\theta)$$

so:

$$\frac{dr}{d\theta} = \frac{k(\cos\theta + m\sin\theta)}{(\sin\theta - m\cos\theta)^2}$$

Therefore:

$$r^2 + \left(\frac{dr}{d\theta}\right)^2 = \frac{k^2}{(\sin\theta - m\cos\theta)^2} + \frac{k^2(\cos\theta + m\sin\theta)^2}{(\sin\theta - m\cos\theta)^4} = \frac{k^2(1+m^2)}{(\sin\theta - m\cos\theta)^4}$$

With these determinations and integrating its first term, equation (71c) may be written as:

$$\Delta t = \frac{l}{c} - \frac{2GM}{c^3}\int_0^\theta \frac{\sin\theta - m\cos\theta}{k}\,\frac{k(1+m^2)^{\frac{1}{2}}}{(\sin\theta - m\cos\theta)^2}\,d\theta$$

$$= \frac{l}{c} - \frac{2GM}{c^3}\int_0^\theta \frac{(1+m^2)^{\frac{1}{2}}}{\sin\theta - m\cos\theta}\,d\theta = \frac{l}{c} - \frac{2GM}{c^3}\int_0^\theta \frac{\sec\theta(1+m^2)^{\frac{1}{2}}}{\tan\theta - m}\,d\theta \tag{73}$$

The first term on the right is the duration of linear travel in the gravity-free vacuum. The second term is the durational difference due to gravity.

Recalling double angle formulae, we may write:

$$\tan\theta = \frac{\sin\theta}{\cos\theta} = \frac{2\sin\frac{\theta}{2}\cos\frac{\theta}{2}}{\cos^2\frac{\theta}{2} - \sin^2\frac{\theta}{2}} = \frac{2\tan\frac{\theta}{2}}{1-\tan^2\frac{\theta}{2}}$$

$$\sec\theta = \frac{1}{\cos\theta} = \frac{1}{\cos^2\frac{\theta}{2} - \sin^2\frac{\theta}{2}} = \frac{\sec^2\frac{\theta}{2}}{1-\tan^2\frac{\theta}{2}}$$

Denoting the second term on the right of equation (73) by $\Delta t'$, substituting from the above equations, and rearrange as:

$$\Delta t' = -\frac{2GM}{c^3}\int_0^{\theta}\frac{m\sec^2\frac{\theta}{2}\,(1+m^2)^{-\frac{1}{2}}}{\left[\left(m\tan\frac{\theta}{2}+1\right)(1+m^2)^{-\frac{1}{2}}\right]^2-1}d\theta$$

Rewrite as:

$$\Delta t' = \frac{2GM}{c^3}\int_0^{\theta}\left[\frac{\frac{1}{2}m\sec^2\frac{\theta}{2}(1+m^2)^{-\frac{1}{2}}}{1+\left(m\tan\frac{\theta}{2}+1\right)(1+m^2)^{-\frac{1}{2}}}+\frac{\frac{1}{2}m\sec^2\frac{\theta}{2}(1+m^2)^{-\frac{1}{2}}}{1-\left(m\tan\frac{\theta}{2}+1\right)(1+m^2)^{-\frac{1}{2}}}\right]d\theta$$

Then, perform the integration:

$$\Delta t' = \frac{2GM}{c^3}\ln\left.\frac{1+\frac{m\tan\frac{\theta}{2}+1}{\sqrt{1+m^2}}}{\left[1-\frac{m\tan\frac{\theta}{2}+1}{\sqrt{1+m^2}}\right]}\right|_0^{\theta} \quad : \frac{m\tan\frac{\theta}{2}+1}{\sqrt{1+m^2}} < 1$$

At the limits of integration:

$$\Delta t' = \frac{2GM}{c^3}\left(\ln\frac{1+\frac{m\tan\frac{\theta}{2}+1}{\sqrt{1+m^2}}}{\left[1-\frac{m\tan\frac{\theta}{2}+1}{\sqrt{1+m^2}}\right]}-\ln\left[\frac{1+\frac{1}{\sqrt{1+m^2}}}{1-\frac{1}{\sqrt{1+m^2}}}\right]\right) \tag{74}$$

Applying half-angle formulae:

$$\tan\frac{\theta}{2}=\frac{\sin\theta}{1+\cos\theta}=\frac{1-\cos\theta}{\sin\theta}=\sqrt{\frac{1-\cos\theta}{1+\cos\theta}}$$

With reference to Figure 3, apply the cosine rule to the triangle ABC:

$$\cos\theta=\frac{r_1^2+r_2^2-l^2}{2r_1r_2}$$

Substituting the above into the previous equation:

$$\tan\frac{\theta}{2}=\left(1-\frac{r_1^2+r_2^2-l^2}{2r_1r_2}\right)^{\frac{1}{2}}\left(1+\frac{r_1^2+r_2^2-l^2}{2r_1r_2}\right)^{-\frac{1}{2}}$$

With equations (72) and the cosine rule applied to triangle ABC:

$$\frac{1}{\sqrt{1+m^2}} = \cos\alpha = -\cos(\pi-\alpha) = \frac{r_2^2 - r_1^2 - l^2}{2r_1 l} \tag{75}$$

Applying the sine rule to triangle ABC and with equations (72):

$$\frac{m}{\sqrt{1+m^2}} = \sin(\pi-\alpha) = \sin\alpha = \frac{r_2}{l}\sin\theta = \frac{r_2}{l}\sqrt{1-\cos^2\theta} = \frac{r_2}{l}\sqrt{1-\left(\frac{r_1^2+r_2^2-l^2}{2r_1r_2}\right)^2}$$

With these determinations we may write:

$$\frac{m\tan\frac{\theta}{2}+1}{\sqrt{1+m^2}} = \frac{r_2}{l}\left[1-\left(\frac{r_1^2+r_2^2-l^2}{2r_1r_2}\right)^2\right]^{\frac{1}{2}}\left[\frac{1-\frac{r_1^2+r_2^2-l^2}{2r_1r_2}}{1+\frac{r_1^2+r_2^2-l^2}{2r_1r_2}}\right]^{\frac{1}{2}} + \frac{r_2^2-r_1^2-l^2}{2r_1 l}$$

$$= \frac{r_2 - r_1}{l} \tag{76}$$

after some reduction. Substituting from equations (75) and (76) into (74) and simplifying:

$$\Delta t' = \frac{2GM}{c^3}\ln\left[\frac{r_1+r_2+l}{r_1+r_2-l}\right]$$

for $r_1 + r_2 > l$. This is the expression for the delay incurred for light's linear travel in the radial gravitational field.

In the case of travel along a radial path, then $r_2 - r_1 = l$. Substituting in the equation above:

$$\Delta t' = \frac{2GM}{c^3}\ln\left(\frac{r_2}{r_1}\right)$$

It is important to keep in mind that r_1, r_2, l, Δt and $\Delta t'$ are coordinate intervals and not local (proper) measurements.

S6. Orbits in Weak Gravity: Relativistic Precession of Mercury's Perihelion

The Schwarzschild solution given in equation (65b) is applicable to orbits around a massive, spherically uniform, non-rotating and uncharged gravitating body that generates a monopole gravitational field as is the case, approximately, in our solar system. Let τ be measured on the orbiting planet - Mercury.

The gravitational force is radial. Without tangential force, the orbit is planar. Since the orbit is coplanar with the solar centre, some simplification is obtained by fixing the coordinate chart so that $\theta = \pi/2$ and $d\theta = 0$. Equation (65b) then becomes:

$$-(cd\tau)^2 = \left(1-\frac{r_s}{r}\right)^{-1}(dr)^2 + r^2(d\phi)^2 - \left(1-\frac{r_s}{r}\right)(cdt)^2 \qquad (65c)$$

The equations of motion of a material point - relatively to the sun a planet approximates such an entity - in a gravitational field may be determined by applying variational techniques. One approach is to apply Hamilton's Principle requiring the vanishing of the variation of the action integral, the integrand of which (in this case) is:

$$S(x^\mu) = -mc^2 d\tau(x^\mu)$$

Here m is the mass of a planet. Let the independent variable q parametrize a set of surfaces intersecting the trajectories, the latter with common fixed end-point events that are labeled 1 and 2. The trajectories may then be considered functions of q and therefore so may the coordinates - x^μ. Now perform the variation:

$$\delta\int_{q_1}^{q_2} S(q)\,dq = -\int_{q_1}^{q_2}\delta\left(mc^2\frac{d\tau(q)}{dq}\right)dq = -\int_{q_1}^{q_2} mc^2\frac{d\delta\tau}{dq}dq = 0 \qquad (77)$$

[See Appendix § A]. Multiplying (65c) by $m/(dq)^2$ and varying with respect to ϕ only (a subset of the variations of the coordinates that may be applied) yield:

$$-m\delta_\phi\left(c\frac{d\tau}{dq}\right)^2 = -2mc^2\frac{d\tau}{dq}\frac{d\delta_\phi\tau}{dq} = 2mr^2\frac{d\phi}{dq}\frac{d\delta\phi}{dq}$$

where δ_ϕ denotes variation with respect to ϕ only. Divide by $2d\tau/dq$:

$$-mc^2\frac{d\delta_\phi(\tau)}{dq} = mr^2\frac{d\phi}{d\tau}\frac{d\delta\phi}{dq}$$

In (77) this implies: $\delta_\phi\int_{q_1}^{q_2} S\,dq = \int_{q_1}^{q_2} mr^2\frac{d\phi}{d\tau}\frac{d\delta\phi}{dq}dq = 0$

Integrating by parts:

$$\delta_\phi\int_{q_1}^{q_2} S\,dq = \left[mr^2\frac{d\phi}{d\tau}\delta\phi\right]\Bigg|_{q_1}^{q_2} - \int_{q_1}^{q_2}\frac{d}{dq}\left(mr^2\frac{d\phi}{d\tau}\right)\delta\phi\,dq = 0$$

The first term on the right vanishes as $\delta\phi$ is zero at the limits of the integration. So for arbitrary $\delta\phi$, identically over the interval between point-events 1 and 2:

$$\frac{d}{dq}\left(mr^2\frac{d\phi}{d\tau}\right) = 0$$

On integration:

$$mr^2\frac{d\phi}{d\tau} \stackrel{\text{def}}{=} L \tag{78}$$

Proceeding similarly with respect to the transitional coordinate t:

$$\delta_t \int_{q_1}^{q_2} S(q)\,dq = \int_{q_1}^{q_2} -mc^2\frac{d\delta_t(\tau)}{dq}dq = \int_{q_1}^{q_2} -mc^2\left(1-\frac{r_s}{r}\right)\frac{dt}{d\tau}\frac{d\delta t}{dq}dq = 0$$

we arrive at:

$$\frac{d}{dq}\left[mc^2\left(1-\frac{r_s}{r}\right)\frac{dt}{d\tau}\right] = 0$$

Then finally:

$$mc^2\left(1-\frac{r_s}{r}\right)\frac{dt}{d\tau} \stackrel{\text{def}}{=} E \tag{79}$$

The constants L and E are the *conserved* parameters of motion in the central gravitational field that are called the *constants of motion*. L is the *angular momentum* and E is the *total energy*. Thus the gravitational field is *conservative*.

Multiplying equation (65c) by $(1 - r_s/r)/(d\tau)^2$ then substituting $d\phi/d\tau$ and $dt/d\tau$ from (78) and (79), respectively, we may write:

$$\left(\frac{dr}{d\tau}\right)^2 = \frac{E^2}{m^2c^2} - \frac{L^2}{m^2r^2}\left(1-\frac{r_s}{r}\right) - \left(1-\frac{r_s}{r}\right)c^2 \tag{80}$$

Now (78) implies:

$$\frac{dr}{d\tau} = \frac{dr}{d\phi}\frac{d\phi}{d\tau} = \frac{dr}{d\phi}\frac{L}{mr^2}$$

Substitute in equation (80) and rewrite as:

$$\frac{d\phi}{dr} = \frac{1}{r^2\sqrt{\frac{1}{B^2} - \left(1-\frac{r_s}{r}\right)\left(\frac{1}{r^2}+\frac{1}{A^2}\right)}}$$

The equation above is the *equation of motion* of a particle in a radial gravitational field where:

$$A = \frac{L}{mc} \quad \text{and} \quad B = \frac{Lc}{E} \tag{81}$$

Substitute $r_s = 2GM/c^2$ in equation (80) and rearrange:

$$\frac{1}{2}\left(\frac{dr}{d\tau}\right)^2 + \frac{L^2}{2m^2r^2} - \frac{GM}{r} - \frac{L^2GM}{m^2c^2r^3} = \frac{E^2}{2m^2c^2} - \frac{c^2}{2} \qquad (82)$$

The first and second terms on the left are the radial and longitudinal components, respectively, of the planet's kinetic energy per unit mass. The third term is Newton's attractive gravitational potential. The last term on the left is an attractive *relativistic* potential term.

Stable circular orbits are possible where the radial component of the velocity is zero and the sum of the energy terms on the left of equation (82) is at a radial stationary point (a radial 'basin of stability'). These considerations yield:

$$\frac{d}{dr}\left(\frac{L^2}{2m^2r^2} - \frac{GM}{r} - \frac{L^2GM}{m^2c^2r^3}\right) = 0$$

Substituting from (81) and $r_s = 2GM/c^2$, this may be written as:

$$\frac{c^2}{2}\frac{d}{dr}\left(\frac{A^2}{r^2} - \frac{r_s}{r} - \frac{A^2r_s}{r^3}\right) = 0$$

This yields:

$$-\frac{2A^2}{r^3} + \frac{r_s}{r^2} + \frac{3A^2r_s}{r^4} = 0$$

Multiplying by r^4/r_s and solving yield the radii of the circular orbits as:

$$r = \frac{A^2}{r_s}\left(1 \pm \sqrt{1 - \frac{3r_s^2}{A^2}}\right) \cong \frac{A^2}{r_s}\left[1 \pm \left(1 - \frac{3r_s^2}{2A^2}\right)\right]$$

Thus, there are two mathematically possible circular orbits of radii:

$$r_{outer} \cong \frac{A^2}{r_s}\left(2 - \frac{3r_s^2}{2A^2}\right) \cong \frac{2A^2}{r_s} \qquad \text{and} \qquad r_{inner} \cong \frac{A^2}{r_s}\left(\frac{3r_s^2}{2A^2}\right) = \frac{3r_s}{2}$$

The approximations are valid as, for Mercury, the value of A is about 9.05 x 10^6 metres and for the sun, r_s is about 2,950 metres.

For our planetary system, the inner orbit is not possible: $3r_s/2$ being much less than the radius of the sun.

Since the gravitational field is conservative then a small perturbation of the circular orbit may lead to oscillations. First, consider orbital oscillations in the science of Newton.

The planet's total energy in Newtonian science, obtained from the expression on the left side of equation (82) by discarding the relativistic fourth term - this action being taken as the equivalent to the 'first approximation' referred to in sub-paragraph d of § K and directly invoked in § S2 - and multiplying by m, is:

$$E' = \frac{m}{2}\left(\frac{dr}{d\tau}\right)^2 + \frac{L^2}{2mr^2} - \frac{GMm}{r} \tag{83}$$

Substituting from equation (78):

$$E' = \frac{m}{2}\left(\frac{dr}{d\tau}\right)^2 + \frac{m}{2}\left(r\frac{d\phi}{d\tau}\right)^2 - \frac{GMm}{r}$$

and since:

$$v^2 = \left(\frac{dr}{d\tau}\right)^2 + \left(r\frac{d\phi}{d\tau}\right)^2$$

we arrive at:

$$E' = \frac{mv^2}{2} - \frac{GMm}{r} \tag{84}$$

where v is the speed. With conservation of energy, oscillations imply alternating transfers of energy between the kinetic and the potential energy components.

Let us consider oscillations of the gravitational potential energy of the planet. The perturbation shifts the planet from the radius of the circle denoted by l in to a radius r_1 or out to a radius r_2. Any one of these shifts initiates a continuing alternation of radial shifts to these same radii. *Assuming* that this motion is a simple harmonic oscillation of the gravitational potential, then:

a. the planet's pre-existent potential in its circular orbit - given by $-GM/l$ - becomes, after perturbation, the *centre of its gravitational potential oscillation.*

b. the alternating shifts in the potential from its centre are equal and opposite:

$$-\frac{GM}{l} - \left(-\frac{GM}{r_1}\right) = -\left[-\frac{GM}{l} - \left(-\frac{GM}{r_2}\right)\right] \tag{85}$$

The magnitude of these shifts is the *amplitude of the oscillation*.

c. dividing the above equation by GM reveals l as the harmonic mean of r_1 and r_2:

$$\frac{1}{r_1} - \frac{1}{l} = \frac{1}{l} - \frac{1}{r_2}$$

d. dividing equation (85) by $(GM)/l$ we get:

$$l/r_1 - 1 = 1 - l/r_2 \stackrel{\text{def}}{=} e$$

which is the *magnitude of the fractional shifts of the gravitational potential.*

e. the oscillation of the potential relative to the centre of potential may be represented in the trigonometric form:

$$-\frac{GM}{r}-\left(-\frac{GM}{l}\right) = -e\frac{GM}{l}\cos\alpha$$

This may be rewritten as:

$$r = \frac{l}{1+e\cos\alpha} \tag{86}$$

Thus at $\alpha = 0$, $r = r_1 = l/(1+e)$ and at $\alpha = \pi$, $r = r_2 = l/(1-e)$. From (86), we see that r_1 and r_2 are the radial extrema. The points $(0, r_1)$ and (π, r_2) are collinear with the origin of the coordinate systems at the centre of the sun, therefore letting the length of the line between the points $(0, r_1)$ and (π, r_2) be 2a, we get:

$$r_1+r_2 = \frac{l}{1+e}+\frac{l}{1-e} = 2a$$

and

$$l = a(1-e^2)$$

so with (86), we get:

$$r = \frac{a(1-e^2)}{1+e\cos\alpha} \tag{86a}$$

This equation describes the stable elliptical orbit first inferred through analysis of empirical astronomical data by Kepler. The angle α is called the *true anomaly*, the quantity e is the *eccentricity* and a is the *semi-major axis*. Now, $r = l$ at $\alpha = \pm\pi/2$. A radius in one of these directions is called the *semi-latus rectum*. (These parameters of the orbital ellipse are discussed in Appendix § B.) The points $(0, r_1)$ and (π, r_2) are called the perihelion and the aphelion, respectively, of the planetary orbit.

Finer analysis of the historical records of astronomical data on Mercury's orbit revealed that the orbit itself rotates (precesses) about its heliocentric focus. After compensating for then known influences, there remained an unexplained precession. That is, Newtonian science was unable to account for this precession.

In considering this precession, the two-fold nature of the planet's movement motivates the application of two frames of reference. One is the heliocentric reference frame F_1 in which the precession is observable. The second frame F_2 is also heliocentric but is referenced to the major axis of the precessing orbital ellipse. Reserving the symbol ϕ for the longitudinal coordinate of F_1, let us denote the true anomaly in F_2 as α. In the two frames, the radial coordinates of the planet are identical. So in frame F_2, the elliptical orbit may be described by equations (86) [24].

Let $u = r_s/r$. The quantity u is directly proportional to the Newtonian gravitational potential. Now we may write:

$$\frac{dr}{d\tau} = \frac{du}{d\phi}\frac{dr}{du}\frac{d\phi}{d\tau}$$

where:

$$\frac{dr}{du} = -\frac{r^2}{r_s}$$

And recalling equation (78) we may write:

$$\frac{d\phi}{d\tau} = \frac{L}{mr^2} = \frac{h}{r^2} \tag{87}$$

where $h \stackrel{\text{def}}{=} L/m$ is the angular momentum per unit mass.

Thus we arrive at:

$$\frac{dr}{d\tau} = -\frac{h}{r_s}\frac{du}{d\phi}$$

Substituting the above equation and $u = r_s/r$ into equation (80) yields:

$$\left(\frac{du}{d\phi}\right)^2 = u^3 - u^2 + \frac{{r_s}^2c^2}{h^2}u + \left(\frac{E^2}{m^2c^2} - c^2\right)\frac{{r_s}^2}{h^2} \tag{88}$$

$$= (u - u_1)(u - u_2)(u - u_3) \tag{89}$$

where the u_j for $j = 1, 2, 3$ are roots of the expression on the right of (88). The retension in (88) of the somewhat modified relativisitic potential term of (82) is the equivalent of the 'second approximation' referred to in sub-paragraph e of § K.

Substitute $u = r_s/r$ into (86):

$$u = r_s(1 + e\cos\alpha)/l \tag{90}$$

Differentiate with respect to α:

$$\frac{du}{d\alpha} = -\frac{r_s e}{l}\sin\alpha \tag{91}$$

This quantity vanishes at the perihelion and at the aphelion in F_2. It follows that in the astronomers' frame F_1, the quantity:

$$\frac{du}{d\phi} = \frac{du}{d\alpha}\frac{d\alpha}{d\phi} \tag{92}$$

also vanishes at these points.

Let $u = u_1$ at $\alpha = 0$ and $u = u_2$ at $\alpha = \pi$. Therefore, equation (90) yields:

$$u_1 = \frac{r_s}{l}(1 + e)$$

and $$u_2 = \frac{r_s}{l}(1-e)$$

Expanding equation (89):

$$\left(\frac{du}{d\phi}\right)^2 = u^3 - (u_1 + u_2 + u_3)\,u^2 + (u_1u_2 + u_2u_3 + u_3u_1)u - u_1u_2u_3$$

Comparing coefficients of the second terms on the right of the above equation and of equation (88) and substituting the expressions for u_1 and u_2 yield:

$$u_3 = 1 - (u_1 + u_2) = 1 - \frac{2r_s}{l}$$

Utilizing equations (90), (91) and (92) and substituting the above expressions for the u_j, we may rewrite equation (89):

$$\left(-\frac{r_s e}{l}\sin\alpha\right)^2\left(\frac{d\alpha}{d\phi}\right)^2 = \left[\frac{r_s}{l}(1+e\cos\alpha) - \frac{r_s}{l}(1+e)\right]\left[\frac{r_s}{l}(1+e\cos\alpha) - \frac{r_s}{l}(1-e)\right]$$
$$\left[\frac{r_s}{l}(1+e\cos\alpha) - 1 + \frac{2r_s}{l}\right]$$

Rewriting:

$$\left(-\frac{r_s e}{l}\sin\alpha\right)^2\left(\frac{d\alpha}{d\phi}\right)^2 = \frac{e^2{r_s}^2}{l^2}(\cos^2\alpha - 1)\left(\frac{3r_s}{l} - 1 + \frac{er_s}{l}\cos\alpha\right)$$

Therefore, simplifying:

$$\left(\frac{d\alpha}{d\phi}\right)^2 = 1 - \frac{3r_s}{l} - \frac{er_s}{l}\cos\alpha$$

Rewrite as:

$$d\phi = \frac{1}{\sqrt{1 - \frac{3r_s}{l} - \frac{er_s}{l}\cos\alpha}}\,d\alpha$$

Applying the binomial expansion then, starting from the perihelion, integrate the above equation around an orbit in F_2:

$$\Delta\phi \cong \int_0^{2\pi}\left(1 + \frac{3r_s}{2l} + \frac{er_s}{2l}\cos\alpha\right)d\alpha$$

where higher order terms in the quantity r_s/l are ignored as this quantity is of the order of 10^{-8}. As α goes from 0 to 2π in F_2, the angle swept in F_1 is:

$$\Delta\phi \cong 2\pi\left(1 + \frac{3r_s}{2l}\right)$$

So the annual precession observable by the astronomer is:

$$\delta\phi \cong \frac{3r_s}{l}\pi$$

Substituting $r_s = 2GM/c^2$ and $l = a(1 - e^2)$:

$$\delta\phi \cong \frac{6\pi GM}{ac^2(1 - e^2)}$$

This is the most common expression of the relativistic precession of a planet's orbit.[25] For Mercury, the relativistic precession amounts to about 43 arc-seconds for each century that passes on the earth. This precession, discovered by the astronomer Le Verrier in 1859, remained inexplicable until 1915 when Einstein accounted for it in his *FGTR* [cf. sub-paragraph e of § K].

♦ ♦ ♦

The equations of the linear elements contain the solutions of the EFE for conditions of induction given in the energy tensors of the material processes. These equations not only describe the resulting field components in their metric tensors but also embody the physics of the inductive material processes in the energy tensors. Thusly, the equations of the linear elements provide significant advances in the unification of the physics of material processes and gravitation.

S7. Strong Gravity: A Note on Schwarzschild Black Holes

As exhausted stars, formerly powered by thermonuclear fusion, shrink under the influence of their own gravity, dependent on the size of their masses, they evolve through various stages. If the process is stopped by the repulsive strong nuclear forces, the collapsed star becomes a neutron star through the forced combination of its protons and electrons to form neutrons. If gravity overcomes the strong nuclear forces then the star shrinks to become a black hole of extreme density. This may happen to stars with masses exceeding that of ten suns. A non-rotating and uncharged black hole is called a Schwarzschild black hole.

In general, the minimum speed at which a projectile can escape a body's gravitational field is called the escape velocity of the gravitating body. This minimum speed of escape occurs when there is mutual cancellation of the kinetic and potential energies. A projectile satisfying this condition will travel to an infinite distant away from the gravitating body without additional energy input or residual energy (apart from the energy of its rest mass).

Recall the modified Schwarzschild solution of equation (82). On the left is the sum of the kinetic and the potential energies. When the total energy E of the projectile is equal merely to its rest mass energy - mc^2 - then the right hand side of this equation vanishes so the kinetic and the potential energies mutually cancel. Now substituting $r_s = 2GM/c^2$ and rewriting the equation:

$$\left(\frac{dr}{d\tau}\right)^2 = -\frac{L^2}{m^2r^2} + \frac{c^2r_s}{r} + \frac{L^2r_s}{m^2r^3}$$

And at $r = r_s$: $$dr/d\tau = c$$

So the escape velocity at the geometrical surface of the sphere at which $r = r_s$ is the speed of light. Notwithstanding the purely mathematical singularity that results in the Schwarzschild solution at the surface of the sphere of radius r_s, at this surface the repulsive centrifugal potential is cancelled by the relativistic potential and only the attractive Newtonian gravitational potential is effective at this surface. Furthermore, as the altitude of the falling body further reduces below a radius of r_s, the attractive relativistic term increases more rapidly than the centrifugal potential term. Therefore, having fell to the surface of the sphere at which $r = r_s$, objects are inexorably drawn downward.

Not even light can escape this fate. Thus the black hole does not emit nor reflect radiation, hence its name. Therefore, the only intrinsic information available about the Schwarzschild black hole, beyond the sphere at which $r = r_s$, is its mass as measured by its gravitational effects. Therefore, the surface of this sphere is called the event horizon.

Clearly, a Schwarzschild black hole is formed only when the radius of the collapsing star becomes less than its Schwarzschild radius.

T. Application of GTR in Global Positioning System Technology

The Global Positioning System (GPS) is a 4-D measurement system based on the use of satellites, clocks, computers, and communications. The basics of GPS and the need for the application of GTR may be summarized as follows:

From above the horizon, four synchronized and adequately separated satellite-based clocks of high precision transmit timing signals and associated 4-D coordinate information of such transmission events. The user has a receiver that has a clock of lower precision, enhanced by a computer. It receives the timing signals and the associated information from the satellites. Based upon transmission intervals, signal speed, satellite positions and a global time keeping reference, the receiver calculates its local position and time.

The clock is the central instrument of the GPS. In general, a clock consists of an oscillator connected to an accumulator. The accumulator counts the cycles. The higher the frequency of the clock (clock rate), the greater is its precision. Stability of clock rate is highly desirable. GTR affirms that the period of an oscillation, measured in coordinate time, depends on the speed and the gravitational potential of the stage of the oscillation. As the frequency is the reciprocal of the period, the clock rate measured in coordinate time also depends on the speed and the potential. This dependence of clock rates, measured in coordinate time, on local conditions is the main reason for the application of GTR in GPS.

Now, all GPS clocks initially have to be calibrated in an Earth-Centered Earth Fixed (ECEF) reference frame and synchronized with standard clocks in the ECEF. An ECEF chart is embedded in a gravitational field and rotates with the earth. However, calculations are simplest if the chart of coordinates is an Earth-Centered Inertial (ECI) frame of reference bereft of non-linear gravitational and motional effects. This is because the kinematics of earth-orbiting satellites and electromagnetic phenomena are most easily described in an ECI frame. The former since earth's motion is ignored. The latter since light travels rectilinearly in an ECI frame. Therefore, there is a need for transformations between the ECEF and the facilitative ECI. Such transformations include relativistic effects due to gravity, earth's rotation, and the motion of satellites and receivers.

The GPS has three segments:

1. The Space Segment nominally consists of twenty-four satellites orbiting the earth twice per day with relative positions such that from any point on earth at least four of these satellites will be above the horizon. These satellites are equipped with high precision atomic clocks. The clocks of this segment transmit timing signals and navigational information to earth-based receivers of users and clocks in the Control Segment of the GPS.

2. The Control Segment (CS) consists of several earth stations with a Master Control station. The high precision atomic clocks in this segment are the standard clocks of the GPS. Stations in the CS monitor the flight of the satellites by radar and receive data from the clocks in the Space Segment. The combined data is sent to the Master Control station where it is processed to provide projected trajectories - ephemerides - of the satellites and data for the correction of the readings of satellite clocks. This information is uploaded to the satellites.

3. The User Segment consists of users with radio receivers equipped with quartz-based clocks of lower precision enhanced by computers. From the satellite clocks, these receivers obtain the timing signals and associated navigational information of the satellites. The local position and time coordinates of the receiver are then determined by the receiver and made available for use.

T1. Synchronization of Standard Clocks in the Control Segment

A convenient approach to the construction of a spherically symmetric chart of the Control Segment may utilize the same approximation of equation (65a) partly by means of which we arrived at equation (71a). Here, however, $ds \neq 0$. So ignoring earth's rotation, the equation of the linear element is: [26]

$$-\,ds^2 = -\,(1 + 2V/c^2)(cdt)^2 + (1 - 2V/c^2)(dr^2 + r^2d\theta^2 + r^2\sin^2\theta d\phi^2) \qquad (93)$$

Here V is a modified Newtonian potential (energy per unit mass) expressed as:

$$V = -\,GM_E(1 + \Delta)/r \qquad (94)$$

Where G is the gravitational constant, M_E is the mass of the earth, and r is the radius from the earth's centre. The number 1 in the brackets expresses the first term of a series in the solution, given in the above equation, to Poisson's form of Newton's law of gravity:

$$\nabla^2 V = 4\pi G\gamma$$

and represents earth's gravitational monopole. (The quantity γ is the mass density of the earth.) Uniform mass distribution makes the perfect sphere the gravitational equivalent of the point-mass - the gravitational *monopole*. Here Δ includes terms reflecting the effects of the oblateness at the poles, the non-uniform distribution of the earth's mass and the movement of the mass of the earth — tectonic plate shifts and diurnal tidal movements. Such departures from sphericity and static uniform mass distribution give rise to a series of gravitational *multipoles*. In GPS calculations, the only gravitational multipole of numerical significance is the first of the series — the *quadrupole*. (There is no gravitational dipole. A dipole requires entities of opposite polarity.)

We transform equation (93) to an ECEF chart by means of the substitutions:

$$t = t' \quad r = r' \quad \theta = \theta' \quad \phi = \phi' + \omega_E t' \qquad (95)$$

Here ω_E is the angular speed of the earth's rotation on its own axis. From this point onwards, we will ignore terms in 1/c of order higher than the second. Equations (93) and (95) yield:

$$
\begin{aligned}
-ds^2 &= -\left(1+\frac{2V}{c^2}\right)(cdt')^2 \\
&\quad + \left(1-\frac{2V}{c^2}\right)\{dr'^2 + r'^2 d\theta'^2 + r'^2 \sin^2\theta' \, [d(\phi' + \omega_E t')]^2\} \\
&= -\left(1+\frac{2V}{c^2} - \frac{1}{c^2} r'^2 \omega_E^2 \sin^2\theta'\right)(cdt')^2 + \frac{2}{c} r'^2 \sin^2\theta' \omega_E \, d\phi' cdt' \\
&\quad + \left(1-\frac{2V}{c^2}\right)(dr'^2 + r'^2 d\theta'^2 + r'^2 \sin^2\theta' \, d\phi'^2)
\end{aligned}
$$

The second term on the right of this equation is due to the *Sagnac effect* exhibited by rotating electromagnetic beams. It is quite small and may be ignored.[27] Thus:

$$
\begin{aligned}
-ds^2 &\cong -\left(1+\frac{2V}{c^2} - \frac{1}{c^2} r'^2 \omega_E^2 \sin^2\theta'\right)(cdt')^2 \\
&\quad + \left(1-\frac{2V}{c^2}\right)(dr'^2 + r'^2 d\theta'^2 + r'^2 \sin^2\theta' \, d\phi'^2) \qquad (96)
\end{aligned}
$$

For a clock fixed on the earth, $dr' = d\theta' = d\,\phi' = 0$. Under these conditions the equation of the linear element becomes:

$$
ds^2 = \left(1+\frac{2V}{c^2} - \frac{1}{c^2} r'^2 \omega_E^2 \sin^2\theta'\right)(cdt')^2 \qquad (97)
$$

Caesium atomic clocks are accurate to about 4 nanoseconds over a day: 5 parts in 10^{14}. There are three factors which make the clock in the ECEF differ from that in the ECI. The relative weights of these factors are as follows: potential of the gravitational monopole — 1, potential due to gravitational multipoles — 1/2000, and the second order Doppler factor due to the Earth's rotation (third term in the first bracket of the last equation) — 1/500. Thus clocks in the ECEF run slower by about 7 parts in 10^{10} compared to clocks in the ECI.[28] Together these influences amount to over 10,000 times the precision of atomic clocks.

Equation (97) may be rewritten as:

$$
\left.
\begin{aligned}
& cd\tau = ds = cdt'\sqrt{1 + 2\Phi/c^2} \\
\text{Where} \qquad & \Phi = V - \frac{1}{2} r'^2 \omega_E^2 \sin^2\theta'
\end{aligned}
\right\} (98)
$$

is the *effective* potential. The effective potential is a negative value that includes both the attractive gravitational potential and an attractive potential term due to earth's rotation - the centripetal potential - in the ECEF. The quantity τ is the *proper time* in seconds measured at rest on the earth, t' is coordinate time (without centripetal or gravitational influences).

We may write: $d\tau = (1 + \Phi/c^2)dt'$

For a clock at rest on the geoid - the equipotential surface of the 'reference ellipsoid'[29] of the same effective gravitational potential Φ_o as that at sea level along the equator – the proper time interval would be:

$$d\tau = (1 + \Phi_o/c^2)dt'$$

Actual standard clocks in the CS are placed at convenience then adjusted to synchronize with the conceptual standard clock in the geoid and with each other.

All standard clocks have the same rate, so coordinate time *may be defined* as:

$$dt'' = d\tau = (1 + \Phi_o/c^2)dt' \qquad (99)$$

Thus coordinate time keeps the same rate as the standard clocks. (Note that 'coordinate time' is purely conceptual. Only a clock at rest and infinitely removed from all matter can measure coordinate time.)

With the *convention* of (99) and substituting from (98), equation (96) of the linear element in the ECEF may be written as:

$$-ds^2 = -\left(1+\frac{2\Phi}{c^2}\right)\left(1+\frac{\Phi_o}{c^2}\right)^{-2}(cdt'')^2 + \left(1-\frac{2V}{c^2}\right)(dr'^2 + r'^2 d\theta'^2 + r'^2 \sin^2\theta'\, d\phi'^2)$$

With the understanding that coordinate time is synchronized with the proper time in the geoid by virtue of the application of (99), we may write the coordinate times without primes. Thus the linear element in the ECEF becomes:

$$-ds^2 = -\left[1+\frac{2(\Phi-\Phi_o)}{c^2}\right](cdt)^2 + \left(1-\frac{2V}{c^2}\right)(dr'^2 + r'^2 d\theta'^2 + r'^2 \sin^2\theta'\, d\phi'^2)$$

T2. Setup and Behaviour of Clocks of the Space Segment

In the non-rotating ECI, the linear element is:

$$-ds^2 = -\left[1+\frac{2(V-\Phi_o)}{c^2}\right](cdt)^2 + \left(1-\frac{2V}{c^2}\right)(dr^2 + r^2 d\theta^2 + r^2 \sin^2\theta\, d\phi^2)$$

This may be rewritten as:

$$-ds^2 = -\left[1+\frac{2(V-\Phi_o)}{c^2} - \left(1-\frac{2V}{c^2}\right)\frac{v^2}{c^2}\right](cdt)^2$$

where $$v = \frac{(dr^2 + r^2 d\theta^2 + r^2 \sin^2\theta\, d\phi^2)^{\frac{1}{2}}}{dt}$$

is the speed of the satellite as measured in the ECI. Furthermore:

$$ds^2 = \left[1 + \frac{2(V - \Phi_o)}{c^2} - \frac{v^2}{c^2}\right](cdt)^2$$

to a first approximation in $1/c^2$. This approximation nullifies, in calculations, the difference in principle between equations (65a) and (93). Taking the square-root:

$$d\tau_S = \frac{ds}{c} = \left[1 + \frac{2(V - \Phi_o)}{c^2} - \frac{v^2}{c^2}\right]^{\frac{1}{2}} dt \cong \left(1 + \frac{V - \Phi_o}{c^2} - \frac{v^2}{2c^2}\right) dt$$

The quantity $d\tau_S$ is the proper time interval of the satellite clock. Rewriting:

$$dt = \left(1 - \frac{V - \Phi_o}{c^2} + \frac{v^2}{2c^2}\right) d\tau_S \qquad (100)$$

The second term in the brackets is a negative potential difference term that contracts the satellite clock's time increment (period), measured in coordinate time, relative to its proper period. As frequency is the reciprocal of the period, this gives rise to a blue-shift of the satellite clock's rate with respect to clocks in the geoid that tick coordinate time. The third term yields STR's time dilation that causes a red-shift of the satellite clock rate measured in coordinate time. (The two effects cancel at a radius of about 9,550 metres,[30] much lower than GPS orbits.)

The influence of the relativistic potential discussed in § S6 is negligible due to the relatively small mass of the earth and the small angular momentum of the satellite. The satellite's orbit is nominally elliptical. As the orbit and the chart are both earth-centered, $dr/d\phi$ vanishes at the radial extrema - apogee and perigee. This implies that at these points:

$$\frac{dr}{d\tau} = \frac{dr}{d\phi}\frac{d\phi}{d\tau} = 0$$

The satellite's total energy in Newtonian science is given by equation (83). Therefore, at the radial extrema, the radial part of the kinetic energy vanishes momentarily, so equation (83), substituting M_E for M, becomes:

$$E' = \frac{L^2}{2mr^2} - \frac{GM_E m}{r} \qquad (101)$$

Since the total energy is conserved we can equate the expressions of the total energy at these radial extrema:

$$E' = \frac{L^2}{2mr_p^2} - \frac{GM_E m}{r_p} = \frac{L^2}{2mr_a^2} - \frac{GM_E m}{r_a} \tag{102}$$

Also, at perigee and apogee $\phi_p = 0$ and $\phi_a = \pi$, respectively, and equation (86a) yields:

$$r_p = a(1 - e) \qquad \text{and} \qquad r_a = a(1 + e) \tag{103}$$

Substituting from equations (103) into (102) and rearranging:

$$\frac{GM_E m^2}{L^2} = \frac{1}{a(1 - e^2)} \tag{104}$$

Since the satellite is negligibly powered and the gravitational field is conservative then the total energy may be arrived at by considering its expression, as given in equation (102), at perigee:

$$E' = \frac{L^2}{2mr_p^2} - \frac{GM_E m}{r_p} = \left(\frac{L^2}{2GM_E m^2 r_p^2} - \frac{1}{r_p}\right) GM_E m \tag{105}$$

Substituting from equations (103) and (104) into (105) and simplifying yield:

$$E' = -\frac{GM_E m}{2a}$$

(The negative sign indicates that the satellite is gravitationally bonded to the earth). Furthermore, equation (84) yields:

$$E' = \frac{mv^2}{2} - \frac{GM_E m}{r} = -\frac{GM_E m}{2a} \tag{106}$$

At the orbital radius of the satellites (about 26,500 kilometres) the gravitational quadrupole is negligible so the gravitational potential approximates:

$$V = -\frac{GM_E}{r} \tag{107}$$

Substituting from equations (106) and (107) into (100) and rearranging we get:

$$dt \cong \left[1 + \frac{\Phi_o}{c^2} + \frac{GM_E}{rc^2} + \frac{GM_E}{c^2}\left(\frac{1}{r} - \frac{1}{2a}\right)\right] d\tau_s$$

We may rewrite this as:

$$dt = \left[1 + \frac{3GM_E}{2ac^2} + \frac{\Phi_o}{c^2} - \frac{2GM_E}{c^2}\left(\frac{1}{a} - \frac{1}{r}\right)\right] d\tau_S \qquad (108)$$

Since $d\tau$ is dependent on the satellite's location in the field and its speed, it is not an absolute differential and its integral is 4-D path-dependent. Integrating equation (108):

$$\int_{path} dt = \int_{path} \left[1 + \frac{\frac{3GM_E}{2a} + \Phi_o}{c^2} - \frac{2GM_E}{c^2}\left(\frac{1}{a} - \frac{1}{r}\right)\right] d\tau_S \qquad (109)$$

On the right, the second term in the square brackets, being negative, gives rise to a constant contraction of the orbiting clock's period of oscillation measured in coordinate time and so constitutes a constant blue-shift of the satellite clock's rate relative to clocks in the geoid that keep coordinate time. Since frequencies are inversely proportional to the time intervals, this shift is compensated by a correction applied to the satellite clock rate *before launch* as:

$$f_{s'} = \left[1 + \frac{\frac{3GM_E}{2a} + \Phi_o}{c^2}\right] f_o \qquad d\tau_{s'} = \frac{f_o}{f_{s'}} d\tau_o \qquad (110)$$

Here, the subscript o denotes the standard clock in the geoid and s′ denotes the clock to be launched. This adjustment is performed on the clock to be placed in orbit with it being at rest relative to a nearby standard clock (the Einstein synchronization). The second term in the square brackets is about -4.4647×10^{-10}, therefore the clock destined for space is made to *run slower*, while at rest in the geoid, than the standard clocks that beat coordinate time.

After the above adjustment and launch, equation (108) is modified to become:

$$dt = \left[1 + \frac{\frac{3GM_E}{2a} + \Phi_o}{c^2} - \frac{2GM_E}{c^2}\left(\frac{1}{a} - \frac{1}{r}\right)\right]\left[1 - \frac{\frac{3GM_E}{2a} + \Phi_o}{c^2}\right] d\tau_o$$

$$\cong \left[1 - \frac{2GM_E}{c^2}\left(\frac{1}{a} - \frac{1}{r}\right)\right] d\tau_S$$

Once in orbit, the clock becomes a standard GPS satellite clock and so τ_o - the standard time interval in the geoid - is replaced by τ_S. Rewriting and integrating:

$$\int_{path} d\tau_S = \int_{path} \left[1 + \frac{2GM_E}{c^2}\left(\frac{1}{a} - \frac{1}{r}\right)\right] dt$$

The second term in the square brackets may be regarded as an error of the orbiting satellite clock's timekeeping with respect to coordinate time as kept by standard clocks in the geoid: *the eccentricity error*. It is so-called as clearly it would vanish if the satellite's orbit was perfectly circular and not elliptical. However, the eccentricity error is a relativistic effect - immune to the accuracy, stability and precision of the satellite clocks – that is due to the variability, under the influence of gravity and motion, of the satellite's proper time with respect to coordinate time. These influences are revealed by rewriting equation (106) as:

$$-\mathrm{GM_E}\left(\frac{1}{\mathrm{a}} - \frac{1}{\mathrm{r}}\right) = -\frac{\mathrm{GM_E}}{\mathrm{r}} + \mathrm{v}^2 \qquad (111)$$

Therefore, the third term in the brackets of equation (109) is, in coordinate time, the correction - Δt_r - of the relativistic eccentricity timing error of the satellite's clock proper time with respect to standard clocks in the geoid.

T3. Determination of the Eccentricity Error in the Control Segment

In the CS, the eccentricity error of a satellite clock is determined by earth stations from radar measurements of the position $\bar{\mathbf{r}}$ and velocity $\bar{\mathbf{v}}$ of the satellite. The monitoring of the eccentricity error is performed in order to provide, in conjunction with timekeeping information garnered from communications with the satellite clock, an assessment of the 'health' of a satellite's clock.

(A number of samples of $\bar{\mathbf{r}}$ and $\bar{\mathbf{v}}$ also permits the inference of a best fit orbit function and the subsequent projection of the ephemeris by Master Control. The information on $\bar{\mathbf{r}}$ and $\bar{\mathbf{v}}$ is also used in flight control.)

To determine the calculation of the eccentricity error in the CS, first consider the Euler formulation of kinematics in two dimensions. Here the radial and polar (longitudinal) unit vectors – $\hat{\boldsymbol{\varepsilon}}_r$ and $\hat{\boldsymbol{\varepsilon}}_\phi$, respectively – may, with respect to a reference axis, be expressed as:

$$\hat{\boldsymbol{\varepsilon}}_r = \mathrm{e}^{i\phi} \qquad \hat{\boldsymbol{\varepsilon}}_\phi = i\hat{\boldsymbol{\varepsilon}}_r = i\mathrm{e}^{i\phi} = \mathrm{e}^{i\left(\phi + \frac{\pi}{2}\right)}$$

Where i is the imaginary unit and e is Euler's number and the base of the natural logarithm.

$$i \stackrel{\text{def}}{=} \sqrt{-1} = \mathrm{e}^{i\frac{\pi}{2}}$$

The transitional derivatives of these unit vectors are:

$$\dot{\hat{\boldsymbol{\varepsilon}}}_r = i\mathrm{e}^{i\phi}\dot{\phi} = \dot{\phi}\hat{\boldsymbol{\varepsilon}}_\phi$$

$$\dot{\hat{\boldsymbol{\varepsilon}}}_\phi = -\,\mathrm{e}^{i\phi}\dot{\phi} = -\,\dot{\phi}\hat{\boldsymbol{\varepsilon}}_r$$

The positional vector is given by:

$$\bar{\mathbf{r}} = r\hat{\boldsymbol{\varepsilon}}_r \tag{112a}$$

Differentiating:

$$\bar{\mathbf{v}} = \dot{\bar{\mathbf{r}}} = \dot{r}\hat{\boldsymbol{\varepsilon}}_r + r\dot{\hat{\boldsymbol{\varepsilon}}}_r = \dot{r}\hat{\boldsymbol{\varepsilon}}_r + r\dot{\phi}\hat{\boldsymbol{\varepsilon}}_\phi \tag{112b}$$

Here $\bar{\mathbf{v}}$ is the satellite's velocity. So we may write:

$$v^2 = \bar{\mathbf{v}}.\bar{\mathbf{v}} = (\dot{r}\hat{\boldsymbol{\varepsilon}}_r + r\dot{\phi}\hat{\boldsymbol{\varepsilon}}_\phi).(\dot{r}\hat{\boldsymbol{\varepsilon}}_r + r\dot{\phi}\hat{\boldsymbol{\varepsilon}}_\phi) = \dot{r}^2 + r^2\dot{\phi}^2 \tag{112c}$$

In a chart with origin at the center of the gravitating body, the force field and the acceleration are radial and the latter is given by:

$$\bar{\mathbf{f}} = f\hat{\boldsymbol{\varepsilon}}_r = \dot{\bar{\mathbf{v}}} = \ddot{\bar{\mathbf{r}}} = \ddot{r}\hat{\boldsymbol{\varepsilon}}_r + 2\dot{r}\dot{\phi}\hat{\boldsymbol{\varepsilon}}_\phi + r\ddot{\phi}\hat{\boldsymbol{\varepsilon}}_\phi - r\dot{\phi}^2\hat{\boldsymbol{\varepsilon}}_r \tag{112d}$$

So equating coefficients of $\hat{\boldsymbol{\varepsilon}}_r$ we get:

$$f = \ddot{r} - r\dot{\phi}^2 \tag{112e}$$

The terms in $\hat{\boldsymbol{\varepsilon}}_\phi$ in equation (112d) mutually cancel identically. This result may also be obtained by differentiating equation (78).

Now as the potential, in the radially-symmetric field, is a function of the radius only then equation (62d) yields the gravitational acceleration as the negative of the radial derivative of the gravitational potential, considering equation (107), we get:

$$f = -\frac{dV}{dr} = -\frac{d}{dr}\left(-\frac{GM_E}{r}\right) = -\frac{GM_E}{r^2} \tag{113}$$

Therefore substituting from equations (112) and (113) into (111):

$$\begin{aligned} GM_E\left(\frac{1}{a} - \frac{1}{r}\right) &= -\left(rf + v^2\right) = -\left[r(\ddot{r} - r\dot{\phi}^2) + \dot{r}^2 + r^2\dot{\phi}^2\right] \\ &= -\left(r\ddot{r} + \dot{r}^2\right) \end{aligned} \tag{114}$$

Now, as the acceleration is radial, equation (114) motivates consideration of the following:

$$\frac{d}{dt}(\bar{\mathbf{v}}.\bar{\mathbf{r}}) = \bar{\mathbf{f}}.\bar{\mathbf{r}} + \bar{\mathbf{v}}.\bar{\mathbf{v}}$$

Recall that the time variable applied to the satellite in (114) is its locally measured time τ. Since the radius of GPS satellites is about 26,500 kilometres and earth's Schwarzschild radius is about 9 millimetres, then $r_S/r \cong 0$. Furthermore, since the velocity of the satellite is much less than the speed of light and with no significant energy apart from its rest mass energy, then $E \cong mc^2$. In light of the foregoing, equation (79) implies $dt \cong d\tau$. With these considerations, substituting from equations (112) into the equation above yields:

$$\frac{d}{d\tau_S}(\bar{\mathbf{v}}.\bar{\mathbf{r}}) = \left(\ddot{r} - r\dot{\phi}^2\right)\hat{\boldsymbol{\varepsilon}}_r . r\hat{\boldsymbol{\varepsilon}}_r + \dot{r}^2 + r^2\dot{\phi}^2 = r\ddot{r} + \dot{r}^2$$

So with (114):

$$-GM_E\left(\frac{1}{a} - \frac{1}{r}\right) = \frac{d}{d\tau_S}(\bar{\mathbf{v}}.\bar{\mathbf{r}})$$

Multiplying by $2/c^2$ and integrating over the path:

$$\Delta t_r = -\int_{path} \frac{2GM_E}{c^2}\left(\frac{1}{a} - \frac{1}{r}\right) d\tau_S = \frac{2\bar{\mathbf{v}}.\bar{\mathbf{r}}}{c^2} \qquad (115)$$

That is, the correction of the eccentricity timing error Δt_r of the satellite clocks performed in the CS may take the form $2\bar{\mathbf{v}}.\bar{\mathbf{r}}/c^2$ exactly.

T4. The Functions of Receivers in the User Segment

From four satellite clocks, the quartz-based clocks of users receive timing signals and navigational information of the satellites (in ephemerides and almanacs uploaded by Master Control to the satellites) that includes coordinate data of the satellites at the times of transmission of the timing signals.

The ephemeris is nominally updated about every thirty seconds. Among the information contained in the ephemeris are data on the following orbital parameters. (Excepting t_{oe}, these are discussed in the Appendix § B):

- e - eccentricity
- a - semi-major axis
- n - mean motion
- ν - true anomaly
- t_{oe} - time of last perigee passage
- M - mean anomaly
- Ω - longitude of ascending node
- ω - argument of perigee
- *i* - angle of inclination of the orbital plane with respect to the equatorial plane

The nominal GPS time of the satellite clock is updated every 1.5 seconds in an ephemeris.

Because of the sensitivity of the compensations to errors in the calculations, GPS has determinate values of several parameters that are to be used by the receivers. These include the following:

- μ - earth's gravitational parameter, $\mu = GM_E = 3.986005\text{x}10^{14}\ m^3s^{-2}$
- c - speed of light, $c = 299{,}792{,}458\ m\ s^{-1}$
- ω_E - earth's rate of rotation, $\omega_E = 7.2921151467 \times 10^{-5}\ rad\ s^{-1}$
- π - 3.1415926535898

The receiver has to capture the satellites' signals, identify the satellites and decode the navigational information. It then performs the timing of the durations of transmission of the timing signals. Next, the receiver performs compensating calculations of the received data to correct the data on the satellites' clocks. Subsequently, the receiver precisely determines the satellites' positions and its position and then calculates its local time.

T4.1 Reception of the Communication Signal

The information broadcast of each satellite is encoded on a carrier of a frequency that is a multiple of its clock frequency. In this code, the bit 1 is represented by a phase reversal and the bit 0 is represented by no phase reversal. In a phase reversal, the electric and the magnetic fields vanish momentarily. Equations (49) reveal that this amounts to the momentary vanishing of the electromagnetic field tensor. As a tensor event, it is recognizable on reception in all frames of reference. Each satellite clock encodes on its carrier a unique code: the coarse acquisition (C/A) code. The C/A codes of all GPS satellites are in the almanac that is broadcast by every satellite for use by receivers to identify satellites.

(Going forward, we restrict the discussion to single frequency - L1 - receivers.)

Apart from identifying the originating satellite, C/A codes are also used in timing the transmission interval. Navigational information is sent along with the C/A code. The navigational information includes data on the position and time of the originating satellite at transmission of the C/A code.

Due to unpredictable frequency shifts of the received signal, the receiver first has to search for and lock on to the signal and then it has to determine the C/A code in order to identify the satellite, perform the timing and decode the navigational data.

T4.1.1 Matching the Frequency Shifts of the Communication Signal

Consider a satellite and a receiver (denoted by the subscript R) in the ECI frame of reference. Since a coordinate time interval is the same everywhere in the ECI, recalling equation (100), we may write:

$$dt = \left(1 - \frac{V_S - \Phi_o}{c^2} + \frac{v_S^2}{2c^2}\right) d\tau_S = \left(1 - \frac{V_R - \Phi_o}{c^2} + \frac{v_R^2}{2c^2}\right) d\tau_R$$

Therefore, we may write:

$$d\tau_R = \left(1 + \frac{V_R - \Phi_o}{c^2} - \frac{v_R^2}{2c^2}\right)\left(1 - \frac{V_S - \Phi_o}{c^2} + \frac{v_S^2}{2c^2}\right) d\tau_S \qquad (116)$$

Recalling equations (107) and (111), we may write:

$$\frac{v_S^2}{2} = -\frac{GM_E}{2a} + \frac{GM_E}{r_S} = \frac{3GM_E}{2a} - \frac{2GM_E}{a} + \frac{GM_E}{r_S}$$

Substituting equation (107) and the above equation into an expanded (116) and rearranging yield:

$$d\tau_R = \left[1 + \frac{V_R - \Phi_o}{c^2} + \frac{\frac{3GM_E}{2a} + \Phi_o}{c^2} - \frac{2GM_E}{c^2}\left(\frac{1}{a} - \frac{1}{r_S}\right) - \frac{v_R^2}{2c^2}\right] d\tau_S$$

Since frequencies are inversely proportional to time intervals, we may write:

$$f_R = \left[1 - \frac{V_R - \Phi_o}{c^2} - \frac{\frac{3GM_E}{2a} + \Phi_o}{c^2} + \frac{2GM_E}{c^2}\left(\frac{1}{a} - \frac{1}{r_S}\right) + \frac{v_R^2}{2c^2}\right] f_s$$

Now applying the correction of (110) and expanding to a first approximation in $1/c^2$, then factoring in the longitudinal Doppler frequency shift, we get:

$$f_R = f_o\left[1 + \frac{v_R^2}{2c^2} - \frac{V_R - \Phi_o}{c^2} + \frac{2GM_E}{c^2}\left(\frac{1}{a} - \frac{1}{r_S}\right)\right] \frac{1 - \frac{\bar{\mathbf{n}}.\bar{\mathbf{v}}_R}{c}}{1 - \frac{\bar{\mathbf{n}}.\bar{\mathbf{v}}_S}{c}} \qquad (117)$$

Here f_R is the centre frequency that the receiver locates and locks on to in order to receive the signal. The unit vector $\bar{\mathbf{n}}$ is in the direction of signal propagation.

In the square brackets, the second term is a fractional relativistic Doppler shift due to the speed of the receiver in the ECI. The third term is the relativistic fractional frequency shift due to the gravitational potential difference between the receiver's location and the geoid. Under usual conditions, these two terms are quite small. The fourth term is a relativistic shift due to the eccentricity of the satellite's orbit. This is the only relativistic shift of the satellite clock's *proper* frequency of transmission relative to that of standard clocks fixed in the geoid.

Consider the eccentricity frequency shifts at the extremal radii $a(1 \pm e)$:

$$\frac{2GM_E}{c^2}\left[\frac{1}{a} - \frac{1}{a(1 \pm e)}\right] = \frac{2GM_E}{ac^2}\left[\frac{(1 \pm e) - 1}{(1 \pm e)}\right] \cong \pm\frac{2GM_E}{ac^2}e \qquad : e \ll 1$$

The semi-major axis is approximately 26.5×10^6 metres and with the values of μ and c, the order of magnitude of the eccentricity frequency shift is $e \times 10^{-10}$.

With regard to the longitudinal Doppler effect, to a first approximation in $1/c$:

$$\frac{1-\frac{\bar{n}.\bar{v}_R}{c}}{1-\frac{\bar{n}.\bar{v}_S}{c}} \cong \left(1-\frac{\bar{n}.\bar{v}_R}{c}\right)\left(1+\frac{\bar{n}.\bar{v}_S}{c}\right) \cong 1-\frac{\bar{n}.(\bar{v}_R-\bar{v}_S)}{c} \cong 1+\frac{\bar{n}.\bar{v}_S}{c} \quad : v_R \ll v_S \ll c$$

The last term is the fractional longitudinal Doppler frequency shift and, with satellite speeds of about 3,900 m/s, is of an order of magnitude of $v_S/c \cong 10^{-5}$.

T4.2 C/A Code Identification and Decoding the Navigational Data

With the frequency locked, the receiver performs correlations of its embedded C/A codes with the broadcast code. This process leads to the recognition of a received C/A code (identification of a satellite) and the matching of its phase.

The information encoded on the satellite is an exclusive OR (XORed) combination of the bits of the navigational data and those of the C/A code. Thus the C/A code of a satellite is required in order to decode its navigational information. In this way, all satellites may use the same bandwidth. This approach to common channel utilization is called code division multiple access.[31]

With the signal's frequency locked and the C/A code and its phase matched (code locked), the navigational data of a satellite is then decoded by XORing the incoming data stream with a locally generated in-phase replica of the C/A code. This deletes the C/A code from the received signal revealing the data.

T4.3 Determination of Pseudorange

To determine the transmission interval, the receiver performs a series of autocorrelations of the carrier-based C/A code with a similarly modulated local replica, the latter signal having a centre frequency of f_R. Each successive autocorrelation runs the replica at an incremented time lag. Measured in the receiver's nominal GPS time frame with respect to the received value of the uncorrected satellite clock's GPS time of C/A code transmission t_j', the time lag that produces the maximal correlation is a first estimate of the transmission interval and determines the phase of the C/A code in the signal being received. (An autocorrelation produces its maximum correlation at zero lag.) The precision of the estimates depends on the magnitude of the increment of the time lag between successive autocorrelations. The autocorrelations may be performed in parallel. The product of this estimate and the speed of light is called the *pseudorange* and is here, for the jth satellite, denoted by ρ_j.

The pseudoranges are inaccurate because they contain errors. These errors include eccentricity timing errors $-\Delta t_{j,r}$ for j = 1, 2, 3, and 4; satellite clock errors

Δt_j; receiver clock bias b_R; and errors due to delays in the ionosphere (due to signal dispersion in the presence of electrons) $\Delta t_{j,iono}$ and in the troposphere (due to refraction in water vapour) $\Delta t_{j,trop}$. (There are other sources of errors that we will not treat with here.) The pseudorange of the jth satellite is given by:

$$\rho_j = c\{[(t_R + b_R) - (t_j + \Delta t_j - \Delta t_{j,r})] + \Delta t_{j,iono} + \Delta t_{j,trop}\} \quad (118)$$

where t_R and t_j are the true times of reception and transmission, respectively. In the first parenthesis is the receiver's uncorrected GPS time of reception of the signal. In the second parenthesis is the satellite clock's uncorrected GPS time of signal transmission - t_j'.

Excepting for the receiver clock bias, the eccentricity timing error and the tropospheric delay, data on the other errors are in the navigational data. The tropospheric delay is estimated by the receiver based on satellite and receiver positions and the use of a model. The clock bias and the eccentricity error are calculated by the receiver.

T4.4 Compensation of the Eccentricity Timing Error by the Receiver

The receiver must calculate and apply the correction of the eccentricity error to the satellite clock's time of transmission of the ephemeris (without data on the satellite's velocity as is available in the Control Segment). This timing error is corrected by the integral in equation (115). This may be rewritten as:

$$\Delta t_r = \int_{path} \left[\frac{2GM_E}{ac^2}\left(\frac{a}{r} - 1\right)\right] d\tau$$

Recall (the discussion on page 76) that under the conditions of GPS orbits $dt \cong d\tau$ and substituting r from equation (B5) in the Appendix we may write:

$$\Delta t_r = \int_{path} \left[\frac{2GM_E}{ac^2}\left(\frac{1}{1 - e\cos E} - 1\right)\right] d\tau = \int_{path} \left[\frac{2GM_E}{ac^2}\left(\frac{e\cos E}{1 - e\cos E}\right)\right] dt$$

Here E is the eccentric anomaly discussed in the Appendix. [Results developed in Appendix § B concerning Keplerian orbital kinematics, transformations between ECI and ECEF frames, and GPS' interpolation strategies will be referenced heavily hereafter.] Substituting dt by means of equation (B14):

$$\Delta t_r = \int_{path} \left[\frac{2GM_E}{ac^2} \left(\frac{e\cos E}{1 - e\cos E} \right) \frac{a^2(1 - e\cos E)}{\sqrt{GM_E a}} \right] dE$$

Simplifying and integrating from last perigee passage:

$$\Delta t_r = \int_{path} \left[\frac{2\sqrt{GM_E a}}{c^2} e\cos E \right] dE = \frac{2\sqrt{GM_E a}}{c^2} e\sin E \tag{119}$$

Given the data in the ephemeris, this expression is used by the receiver to calculate the correction of the timing delay due to the eccentricity of the satellite's orbit.

Firstly, the receiver estimates the time elapsed between the last perigee passage t_{oe} and the best available estimate of the time of transmission of the timing signal:

$$t_k = t'_j - \Delta t_j - t_{oe}$$

Then, given the value of e and data on the mean anomaly M and the mean motion n in the ephemeris, E is iteratively calculated in the receiver by means of (B17). For example, the following iteration may be used:

$$E_i = M + e\sin(E_{i-1}) \qquad M = nt_k \qquad E_0 = 0 \qquad : i = 1, 2, 3, \ldots$$

For the range of values of the eccentricity common in GPS (< 0.02), the iteration is completed in a few cycles. Then equation (119) may be evaluated.

With the semi-major axis being approximately 26.5 x 10^6 metres and with the values of μ and c, equation (119) yields the magnitude of order of the eccentricity error and of its correction as e x 10^{-6}.

T4.5 Determination of Local Time and Position

The most accurate estimates of the times of C/A code transmissions are then determined by the following:

$$t_j = t'_j + \Delta t_{j,r} - \Delta t_j \qquad : j = 1, 2, 3, 4$$

The value of E as determined above and data in the ephemeris on the semi-major axis and the eccentricity are processed by the receiver to produce the orbital radius and the true anomaly of the satellite by means of equations (B5), (B8), and (B9) of the Appendix. With data on the argument of perigee, these polar coordinates are converted, by equations (B18), to Cartesian coordinates in the orbital plane. With data on the angle of inclination and the longitude of the ascending node, the position of the satellite is then transformed by equations (B19) into GPS' ECEF Cartesian reference system.[32] Subsequent transformation by means of equations (B20) yields the satellite's positional 3-vector $\bar{\mathbf{r}}_j$ in an ECI coordinate system.[33]

The receiver's ECI coordinates at signal reception ($\bar{\mathbf{r}}$, t_R) may then be determined by simultaneously solving the following four equations:

$$c^2(t_R - t_j)^2 = |\bar{\mathbf{r}} - \bar{\mathbf{r}}_j|^2 \qquad ;j = 1, 2, 3, 4 \quad (120)$$

(Accommodation of different signal reception times and of the movement of the receiver during signal transmission is not discussed.)

The time of signal reception t_R is available only after a time lag determined mainly by the speed of solving equations (120). So it may be difficult to use its value to correct the receiver's nominal GPS time. In order to correct the receiver's time, combine equations (118) and (120) to yield:

$$b_R = \frac{1}{c}\left(\rho_j - |\bar{\mathbf{r}} - \bar{\mathbf{r}}_j|\right) + \Delta t_j - \Delta t_{j,r} - \Delta t_{j,iono} - \Delta t_{j,trop} \qquad (121)$$

Thus, having solved equations (120), the receiver calculates its clock bias and corrects its GPS time by conveniently subtracting its clock bias from its current nominal GPS time.

Local time is calculated from GPS time by way of Universal Coordinated Time (UTC).[34] GPS time is an offset from UTC and local time has an offset from UTC corresponding to the longitude of its locality. The offset of GPS time from UTC is contained in the almanac broadcast by the satellite in the navigational data. The data on the offset of local time from UTC is embedded in the receiver.

The receiver's ECI space coordinates may be transformed, firstly to GPS' ECEF by applying equations (B21), then to latitude, longitude and altitude (LLA).[35]

The calculation of the couple ($\bar{\mathbf{r}}$, t_R) performed by the receiver using equations (120) is based on the constant speed of light in an ECI frame in the vacuum.[36] Constancy of the speed of light in the ECI is the *basic* principle of GPS.

T5. A Potential Improvement in the Accuracy of the GPS

The extended duration of travel of the satellite's signal in the gravitational field gives rise to an additional source of error in the pseudorange. Currently, this error is not corrected. This error may be corrected by suitably applying the results given in § S5.2.

Part II

The Illusions of Space, Time, and Spacetime

Part 2

The Illusions of Space, Time and Spacetime

"That this requirement of general covariance, which takes away from space and time the last remnant of physical objectivity, is a natural one… "[2]

Time – through which passes the perceived event -
Seems quite profound, though Space seems evident.
Both are alike; twins from the same parent.
Space seems simpler since air is transparent.

Space and Time are cultured orders of things
So descriptions in both the same word brings.
Towards, Before, and After do not tell
Whether in Time or in Space that they dwell.

Unchanging Space and changing Time we glean.
Yet, in practice, lone Distance-Past is seen:
Universal Light that takes There to Here
Has its traveling Time, Science made clear.

Light illuminates the Distance-Past's sphere.
Inferred Future is neither Here nor There.
Fleet-footed Present is to Here stuck fast
And all that is seen was There in the Past.

Thought, for mundane life, conceives Time and Space.
But within both we perceive each thing's face.
Spacetime is Siamese ideality
Science found out in Relativity

Space, Time, and Spacetime are forms our systems
Of reference take to describe items
That are materially existent
Using rod, clock, or other instrument.

Systems of reference are factitious
Einstein wrote and he was never fractious,
But in the interests of great Science,
He exposed these myths of experience.

Still there remains to reveal the bases
Of these longstanding illusive cases.
For, as we know, illusions are quite real,
But mistaken, true contents they conceal.

Not considering Relativity,
Space is an emptied contiguity.
Co-existence is clearly the basis
Of such absolute side-by-sidedness.

For Time in mind to arise term-attired,
Simultaneity is required,
And so is its opposite: succession.
On duality rests this conception.

The Observed and the Observer must be
Of a light-based simultaneity.
Yet, Observed must include for Observer
A succession: Time's Before and After.

It is Time that gives order to Being,
Not speaking of the Space that things are in,
But to say what came first and what next.
Time forms as things relatively exist.

Measurement implies a notion clearer.
One process' lifetime is measured ever
By another's - nuclear, solar, or clock:
There's no Time without the Other's "tick-tock".

Void lifetime - incessant repetition,
Cycles of emergence/expiration,
The only change is that of succession -
That's the chronometer's criterion.

Natural chronometers, recognized
By the coincidences they realized,
Gave rise to a non-spatial reference
For an order of things in Existence.

Then, finding many a natural clock,
Mental leap gave nascent concept its frock.
Chronometers to Time: pure abstraction;
Counting cycles quantifies duration.

With terms from spatial order to borrow
And some created; Yesterday, Now, Tomorrow,
Since then, those things that do not co-exist,
With no shared Space, are ordered on Time's List.

Spacetime is cosmic ideology
Of much use, but remains a parody
Of empirical relationships cut
From things that exist, then beside them put.

Spacetime extends Cartesian spatial grid
So the Hand-Maiden may do Science's bid.
To impose was never Her intention:
Elapsed Time is not the fourth dimension.

The most basic relationships of things
Arise from things merely being beings.
These are Co- and Relative Existence;
Space and Time are their forms in Pre-Science.

Appendix

A. A Note on the Variation of Coordinate Derivatives

Consider the points $Q(x_1, x_2, \ldots, x_n)$ and $Q_1(x_1+dx_1, x_2+dx_2, \ldots, x_n+dx_n)$ located in an n-dimensional domain. For brevity, we shall denote these points as $Q(x_i)$ and $Q_1(x_i+dx_i)$, for $i = 1, 2, \ldots, n$. (For illustrative purposes, imagine these points to be on a surface). As a result of an infinitesimal variation (say, a smooth infinitesimal deformation of the surface) denoted by the operator δ, the points Q and Q_1 become $Q'(x_i+\delta x_i)$ and $Q'_1((x_i+dx_i) + \delta(x_i+dx_i))$, respectively. As such, the differences of the coordinates of Q' and Q'_1 are given by:

$$dx_i + \delta dx_i \qquad : i = 1, 2, \ldots, n$$

Alternatively, the point Q'_1 may be considered as being produced from Q' by infinitesimal differential changes of the latter's coordinates, whereby we may express the former as $Q'_1((x_i+\delta x_i) + d(x_i+\delta x_i))$. In which case, the differences of the coordinates are now given by:

$$d(x_i + \delta x_i) \qquad : i = 1, 2, \ldots, n$$

Equating these expressions of the differences yields $\delta dx_i = d\delta x_i$.

Dividing this equation by $d\lambda$, where λ is an independent variable that does not participate in the variation, we get:

$$\delta\left(\frac{dx_i}{d\lambda}\right) = \frac{d\delta x_i}{d\lambda}.$$

B. Parameters and Reference Frames of the Elliptical Orbit

B.1 Keplerian Orbital Kinematics

The ellipse shown below is circumscribed by a circle of radius a - called the auxiliary circle - and inscribed by a circle of radius b - called the minor auxiliary circle. In the Cartesian frame shown, its equation is given by:

$$\frac{x^2}{a^2} + \frac{y^2}{b^2} = 1 \qquad \text{(B1)}$$

The quantities a and b are the lengths of *the semi-major axis* and *the semi-minor axis*, respectively. The ellipse has a parameter called the *eccentricity* denoted by the letter e. It is defined as:

$$e \stackrel{\text{def}}{=} \sqrt{1 - \frac{b^2}{a^2}} \qquad \text{(B2)}$$

For a = b, the eccentricity is zero and the ellipse becomes a circle.

In orbital physics, there is a parameter called the *eccentric anomaly* denoted by the letter E. It parametrizes the position of a satellite at the point (x, y) on the orbital ellipse and is given in the equations:

$$\cos E = \frac{x}{a} \qquad \text{(B3)}$$

$$\sin E = \frac{y}{b} \qquad \text{(B4)}$$

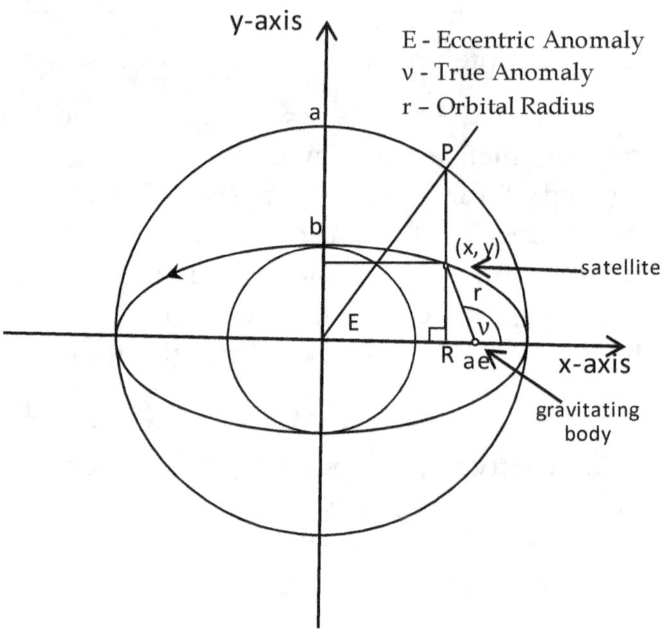

Fig. 1. Parameters of the Orbital Ellipse

Geometrically, this is described by a line drawn from the origin at an angle E to the positive x-axis and intersecting the auxiliary circle at a point P that is vertically above the point (x, y). Note that equations (B3) and (B4) substituted in (B1) transform the equation of the ellipse into the Pythagorean trigonometric identity.

The ellipse is related to two points on its major axis (here the latter is colinear with the x-axis from x = −a to x = a) at distances ±ae from the centre of the ellipse. These points are called the *foci* of the ellipse.

A satellite, natural or artificial, generally describes an elliptical orbit. This is Kepler's first law. In elliptical orbits, the massive gravitating body is located at a focus. The nearest point of the orbit to the gravitating body is called the *periapsis*. The angle swept by the orbital radius since periapsis passage is called the *true anomaly* and is here denoted by ν.

In Figure 1, given the geometry of the triangle with apices at the centre of the gravitating body, the satellite's position (x,y) and the point R we may apply Pythagoras' theorem yielding:

$$r^2 = y^2 + (ae - x)^2$$

Substituting from equations (B3) and (B4) and invoking the Pythagorean identity we first get:

$$r^2 = b^2(1 - \cos^2 E) + (ae - a\cos E)^2$$

Now substituting b from equation (B2), expanding and simplifying yield:

$$r^2 = a^2(1 - 2e\cos E + e^2\cos^2 E) = a^2(1 - e\cos E)^2$$

And the *orbital radius* may be expressed as:

$$r = a(1 - e\cos E) \tag{B5}$$

The x-coordinate of the satellite may be expressed as:

$$x = ae + r\cos\nu$$

Subsituting this equation and (B5) into equation (B3):

$$\cos E = \frac{ae + a(1 - e\cos E)\cos\nu}{a}$$

Rewriting this equation we may express the cosine of the eccentric anomaly in terms of the true anomaly as:

$$\cos E = \frac{e + \cos\nu}{1 + e\cos\nu} \tag{B6}$$

By means of the Pythagorean identity we may also express the sine of the eccentric anomaly in terms of the true anomaly.

$$\sin^2 E = 1 - \cos^2 E = 1 - \left(\frac{e + \cos\nu}{1 + e\cos\nu}\right)^2$$

Rewriting the equation above as:

$$\sin^2 E = \frac{(1 - e^2)\sin^2\nu}{(1 + e\cos\nu)^2}$$

Finally:

$$\sin E = \frac{\sqrt{1 - e^2}\sin\nu}{1 + e\cos\nu} \tag{B7}$$

By rearranging equation (B6), we may express the cosine of the true anomaly in terms of the eccentric anomaly:

$$\cos\nu = \frac{\cos E - e}{1 - e\cos E} \tag{B8}$$

And its sine also, by means of the Pythagorean identity:

$$\sin^2\nu = 1 - \left(\frac{\cos E - e}{1 - e\cos E}\right)^2$$

yeilding:

$$\sin\nu = \frac{\sqrt{(1 - e^2)}\sin E}{1 - e\cos E} \tag{B9}$$

Substituting equation (B6) in (B5) yields an expression for the orbital radius in terms of the true anomaly:

$$r = a\left[1 - e\left(\frac{e + \cos\nu}{1 + e\cos\nu}\right)\right] = \frac{a(1 - e^2)}{1 + e\cos\nu} \tag{B10}$$

The closed figure in the illustration in Figure 2 approximates an infinitesimal right-angled triangle. As such, its enclosed area is approximately given by:

$$dA = r^2 d\nu/2$$

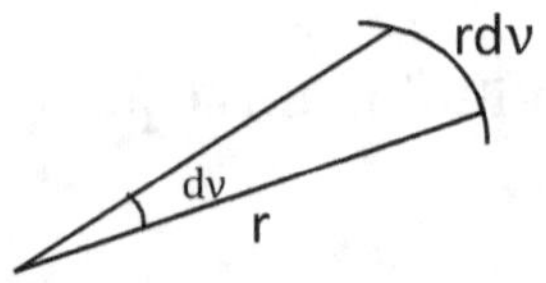

Fig. 2. Infinitesimal Area Swept by Orbital Radius

The rate that the area is being swept by the orbital radius is given by:

$$\frac{dA}{dt} = \frac{r^2}{2}\frac{d\nu}{dt}$$

Invoking equation (87) of § S6 and identifying ϕ with ν and $d\tau$ with dt (see discussion on the latter identification in § T2 on page 76) we may write:

$$\frac{dA}{dt} = \frac{r^2}{2}\frac{d\nu}{dt} = \frac{h}{2} \tag{B11}$$

where h is the *angular momentum per unit mass* of the satellite and is an orbital constant [cf. equations (78) and (87) in § S6]. Equation (B11) expresses Kepler's second law: equal areas are swept by the orbital radius in equal times.

Now differentiating equation (B8) with respect to E:

$$(-\sin\nu)\frac{d\nu}{dE} = \frac{-\sin E(1 - e\cos E) - (\cos E - e)e\sin E}{(1 - e\cos E)^2}$$

Simplifying and substituting from equation (B9) on the left of the equation above:

$$\left(-\frac{\sqrt{1 - e^2}\sin E}{1 - e\cos E}\right)\frac{d\nu}{dE} = -\frac{(1 - e^2)\sin E}{(1 - e\cos E)^2}$$

And

$$\frac{d\nu}{dE} = \frac{\sqrt{1 - e^2}}{1 - e\cos E}$$

Substituting dν by means of equation (B11):

$$dt = \frac{r^2}{h}\frac{\sqrt{1 - e^2}}{(1 - e\cos E)}dE \tag{B12}$$

Equations (87) and (104) yield:

$$h = \frac{L}{m} = \sqrt{GM'a(1 - e^2)} \qquad (B13)$$

Where the gravitating mass is now denoted as M′. Substituting r from equation (B5) and h from equation (B13) into equation (B12) yields:

$$dt = \frac{a^2(1 - e\cos E)^2}{\sqrt{GM'a(1 - \epsilon^2)}} \frac{\sqrt{(1 - e^2)}}{(1 - e\cos E)} dE$$

$$= \frac{a^2(1 - e\cos E)}{\sqrt{GM'a}} dE \qquad (B14)$$

Integrating from t = 0 at periapsis passage:

$$\int_0^t dt = \int_0^E \frac{a^2(1 - e\cos E)}{\sqrt{GM'a}} dE$$

$$t = \frac{a^2(E - e\sin E)}{\sqrt{GM'a}} \qquad (B15)$$

We may rewrite this as:

$$t\sqrt{GM'/a^3} = nt = E - e\sin E \stackrel{\text{def}}{=} M \qquad (B16)$$

where M is called the *mean anomaly* and $n \stackrel{\text{def}}{=} \sqrt{GM'/a^3} = M/t$ is termed the*mean motion*. Equations (B16) are forms of Kepler's equation.

From equations (B12), (B14), (B2) and (B17) we may write:

$$\frac{dA}{dt} = \frac{h}{2} = \frac{\sqrt{GM'a(1 - e^2)}}{2} = \frac{b}{a}\frac{\sqrt{GM'a}}{2} = \frac{ab\sqrt{GM'/a^3}}{2} = \frac{ab}{2}n$$

Now the area of the ellipse $A_e = \pi ab$, so we may usefully write:

$$n = \frac{2\pi}{\pi ab}\frac{dA}{dt} = 2\pi\frac{d}{dt}(A/A_e) = \sqrt{GM'/a^3}$$

Therefore the mean motion n is the rate of change of the swept fractional area of the ellipse scaled by 2π. It is constant in accord with Kepler's second law.

Integrating the equation above, measuring both the duration and the finite swept area A from periapsis passage, and invoking (B17) we have:

$$M = nt = 2\pi(A/A_e)$$

Clearly, the mean anomaly M is the fractional area of the ellipse swept by the orbital radius since periapsis passage and scaled by 2π.

B.2 Orbital Reference Frames

From Figure 1, it is clear that in a Cartesian coordinate system with origin at the centre of the gravitating body, with its x′-y′ plane coplanar with that of the orbit and with its positive x′-axis going through the periapsis, the satellite's coordinates are given by the following transformation:

$$x' = r\cos\nu \quad \text{and} \quad y' = r\sin\nu \tag{B17}$$

Figure 3 shows an orbit in relationship to an earth-centered Cartesian reference system, the latter with its x-y plane coplanar with the equatorial plane. Let the x-axis be aligned with the reference meridian.

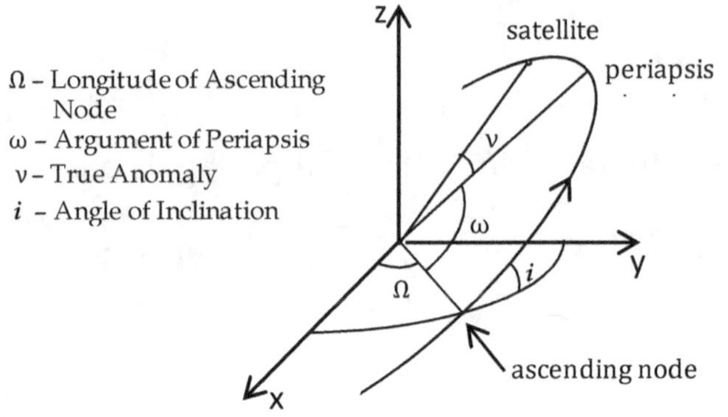

Fig. 3. Angular Relationships of the Orbital Ellipse and the Coordinate System in the Equatorial Plane

Let i be the *angle of inclination* of the orbital plane with respect to the equatorial plane. The *ascending node* is the point of intersection of the orbit with the equatorial plane as the satellite goes above the latter.

If the axes of the coordinate system that is the frame of reference of equations (B17) are rotated so that the positive x′-axis goes through the *ascending node*, then these equations become:

$$x' = r\cos\Phi \quad \text{and} \quad y' = r\sin\Phi \tag{B18}$$

where $\Phi = \omega + \nu$ is the *argument of latitude*.

Figure 4 shows the equatorial plane of Figure 3 with the x′-y′ reference frame of equations (B18) and the satellite's position projected normally onto it. Intervals in the direction of the y′-axis are thereby foreshortened in their projections by a factor of cos*i*.

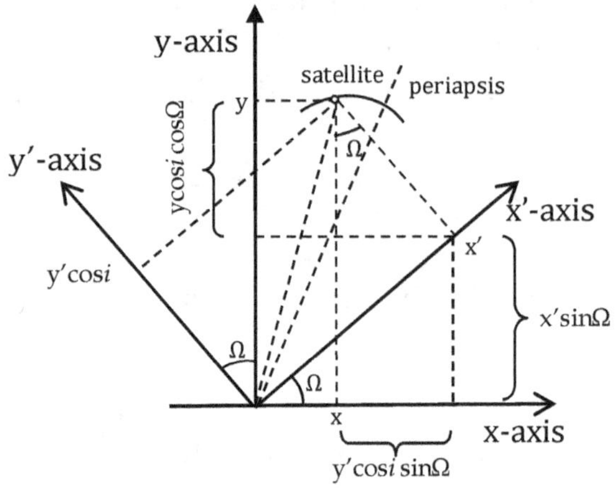

Fig. 4. Orbit-Based Reference System Projected onto the ECEF's Equatorial Plane

Transforming equations (B18) to an earth-centered and earth-fixed (ECEF) Cartesian frame with its x-y plane coplanar with the equatorial plane, we get:

$$\left.\begin{aligned} x &= x'\cos\Omega - y'\cos i\,\sin\Omega \\ y &= x'\sin\Omega + y'\cos i\,\cos\Omega \\ z &= y'\sin i \end{aligned}\right\}\text{(B19a)}$$

In general, due to perturbations of the orbit - caused by gravity (due to earth's rotating, non-spherical and dynamic mass distribution and solar and lunar fields), solar wind, and satellite jets - the angle of inclination and the *longitude of the ascending node* will be changing. With these considerations, the transformation to an ECEF reference system will require modifications.

If the angles are referenced to an initial time t_o and their rates of change are constant (or suitably averaged), then the following equations augment equations (B19a) in constituting the transformation to the ECEF:

$$\left.\begin{aligned} i &= i_o + \Delta t\frac{di}{dt} \\ \Omega &= \Omega_o + (\dot{\Omega} - \omega_E)\Delta t \\ \Delta t &= t - t_o \\ i_o &= i(t_o) \\ \Omega_o &= \Omega(t_o) \end{aligned}\right\}\text{(B19b)}$$

where $\omega_E = 7.2921151467 \times 10^{-5}$ rad/sec is the *rate of earth's rotation* and $\dot{\Omega}$ is the rate of change of the longitude of the ascending node (*rate of right ascension*).

The ECEF coordinates produced by equations (B19) may be transformed to an earth-centered inertial (ECI) Cartesian coordinate system with its x″-y″ plane coplanar with the equatorial plane. Generally, the transformation may be performed at a time t subsequent to a time t_o of the instant of establishment of the coincidence of the ECI frame with that of the ECEF by rotation of the latter around the common z-axis (Figure 5) yielding:

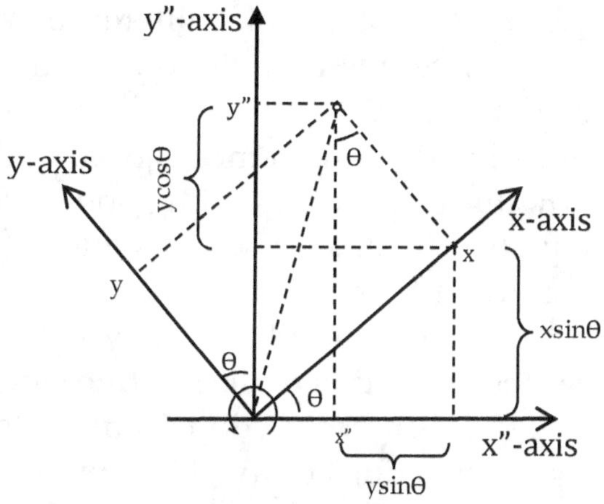

Fig. 5. ECEF to ECI Transformation

$$\left.\begin{aligned} x'' &= x\cos\theta - y\sin\theta \\ y'' &= x\sin\theta + y\cos\theta \\ z'' &= z \\ \theta &= \omega_E(t - t_o) \end{aligned}\right\}\text{(B20)}$$

The transformations of equations (B18), (B19) and (B20) are found in the Global Positioning System. They are specified to be performed by GPS receivers (cf. § T4.5).

An inverse transform of (B20) must also be used by GPS receivers. It may be derived quite easily from the same equations. It comprises the following:

$$x = x''\cos\theta + y''\sin\theta \qquad y = y''\cos\theta - x''\sin\theta \qquad z = z'' \quad \theta = \omega_E(t - t_o) \qquad \text{(B21)}$$

B.3 On the Interpolation Strategies of the GPS

In GPS, the values of the elements given in the navigational data are normally updated periodically. In order to obtain useful values of the elements in between these cutovers, interpolation methods have to be applied by the receivers of users. Interpolation, as opposed to extrapolation, is used because the data transmitted by the satellite to the receivers are normally referenced to a future time. In general, the orbits are subject to certain influences that make them differ from ideal Kleperian trajectories. A specific interpolation strategy is applied depending on the characteristics of the influences on a specific element.

For any element y(t) with aperiodic influences an interpolation polynomial (based on the Taylor series) may be used:

$$y = a_0 + a_1\Delta t + a_2(\Delta t)^2 + a_3(\Delta t)^3 + \cdots + a_N(\Delta t)^N \qquad \text{(B22)}$$

Where Δt is the time difference with respect to the reference time t_o of the interpolation,[37] a_o is a correction offset, and the a_k for $k > 0$ are correction coefficients. A linear interpolation strategy is applied to the estimation of the angle of inclination, the semi-major axis, the mean anomaly, the mean motion and the longitude of the ascending node. Here, only the first two terms on the right of equation (B22) are used. Equations (B19b) are examples. A quadratic polynomial including terms such as the first three of (B22) is specified for the correction of satellite clock time.[38] A relativistic term calculated by the receiver is also required for the correction of the satellite clock time (cf. § T3.4).

Offsets and correction coefficients are determined by projections based on the use of functions derived by application of least square fit algorithms to recently established time-stamped data of the parameters.

For elements with periodic influences, trigonometric interpolation polynomials (based on the Fourier series) may be used:

$$y = b_o + b_1 \sin(x) + b_2 \cos(x) + b_3 \sin(2x) + b_4 \cos(2x) + \cdots + b_N \cos\left(\frac{Nx}{2}\right) \tag{B23}$$

here x is periodic.[39] The quantity b_o is a correction offset and the b_j for $j > 0$ are correction coefficients.

The major source of periodic influences is the quadrupole gravitational potential term (cf. § T1 page 68). (The GPS' reference model is based on a gravitational monopole.) This potential term may be expressed as: [40]

$$\frac{GM_E}{2r} J_2 \left(\frac{a_1}{r}\right)^2 (3\cos^2\theta - 1) \tag{B24}$$

Here G is the gravitational constant, M_E is earth's mass, $J_2 = 1.0826300 \times 10^{-3}$ is the earth's quadrupole moment coefficient, a_1 is earth's equatorial radius, and r and θ are, respectively, the radial and the colatitudinal coordinates of an earth-centered spherically-symmetric reference frame.

Twice daily, the satellites execute orbits inclined at about 55° with proportionately small radial changes (eccentricities < 0.02). So the potential of the gravitational quadrupole is modulated mainly by an oscillating colatitudinal coordinate that varies from 35° to 145° at a frequency of two cycles per day.

In GPS, the satellite's position is most sensitive to periodic influences of the qradrupole potential and, therefore, so are its elements - the argument of latitude, the angle of inclination and the radius. Although the radius is not directly given in the navigational data, a first approximation is obtained by applying equation (B5) with the value of the eccentricity specified in the navigational information and estimates of the semi-major axis and the eccentric anomaly made by the receiver based on navigational data and then the specified correction is applied.

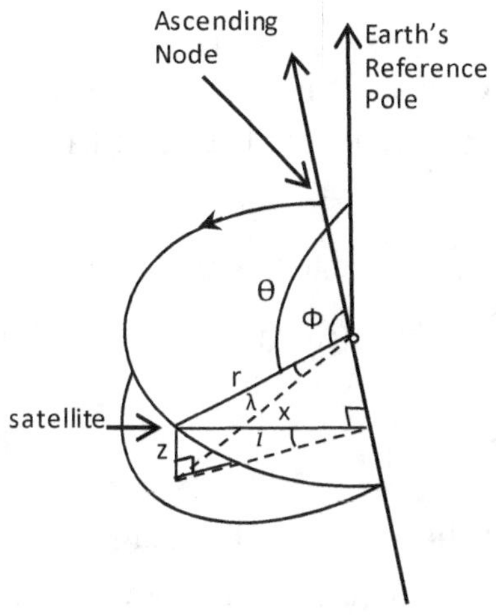

Fig. 6. Projection of Certain Elements of the Orbit onto the Equatorial Plane

From the geometry of Figure 6, the following relationships are evident:

$$\lambda = \pi/2 - \theta \quad \sin\lambda = z/r \quad \sin i = z/x \quad \sin\Phi = x/r$$

So: $\sin\lambda = \sin(\pi/2 - \theta) = \cos\theta = \sin i \sin\Phi$ (B25)

Now: $\sin^2\Phi = (1 - \cos 2\Phi)/2$

Squaring equation (B25), multiplying by three and substituting from the above equation:

$$3\cos^2\theta = 3\sin^2 i\,(1 - \cos 2\Phi)/2$$

For $i \cong 55^\circ$: $$\frac{3}{2}\sin^2 i \cong 1$$

Therefore: $$3\cos^2\theta - 1 \cong -\cos 2\Phi$$

Substitution into equation (B24) yields:

$$-\frac{GM_E}{2r} J_2 \left(\frac{a_1}{r}\right)^2 \cos 2\Phi$$

This reveals that, for GPS satellites, the gravitational quadrupole potential may be approximately expressed in the form of a purely sinusoidal function with an argument of twice the argument of latitude and a frequency of four cycles per day.

Therefore, the interpolation strategies applied in the correction of the argument of latitude, the angle of inclination and the radius may usefully take the following truncated form of (B23):

$$y = c_s \sin 2\Phi + c_c \cos 2\Phi$$

where Φ is the uncorrected argument of latitude and c_s and c_c are the correcting coefficients all given in the navigational data.

C. Alternative Derivation of the Einstein Field Equations

Consider the action integral over the 4-volume τ:

$$\int_\tau \left(\mathcal{L} - \frac{\mathfrak{R}}{2\kappa}\right)\sqrt{-g}\,d\tau \tag{C1}$$

where $\mathcal{L}$ is the Lagrange density of matter (integrated over the translational coordinates it becomes the Lagrangian), $\mathfrak{R}$ is the scalar of the Riemann-Christoffel tensor and the constant $\kappa = 8\pi G/c^4$, with G being the gravitational constant.

Now recall that:

$$\sqrt{-g'}\,d\tau' = \sqrt{-g}\,d\tau$$

where the prime denotes a coordinate system transform [cf. § D]. The product $\sqrt{-g}\,d\tau$ is therefore invariant under coordinate transformations. Now in a coordinate system where $\sqrt{-g} = 1$, this product becomes equal to the infinitesimal coordinate volume $d\tau$. Thus the 4-volume $dV = \sqrt{-g}\,d\tau$ may be considered as the invariant *proper* volume corresponding to the coordinate volume element $d\tau$. Hence, the integral (C1) is invariant under transformation.

Applying Hamilton's principle:

$$\delta \int_\tau \left(\mathcal{L} - \frac{\mathcal{R}}{2\kappa}\right)\sqrt{-g}\,d\tau = 0 \tag{C2}$$

Performing the variation:

$$\int_\tau \left[\sqrt{-g}\left(\delta\mathcal{L} - \frac{\delta\mathcal{R}}{2\kappa}\right) + \left(\mathcal{L} - \frac{\mathcal{R}}{2\kappa}\right)\delta\sqrt{-g}\right] d\tau = 0 \tag{C3}$$

Since: $$\mathcal{R} = g^{\mu\nu}\mathcal{R}_{\mu\nu}$$

where $\mathcal{R}_{\mu\nu}$ is the contracted Riemann-Christoffel tensor, therefore:

$$\delta\mathcal{R} = g^{\mu\nu}\delta\mathcal{R}_{\mu\nu} + \mathcal{R}_{\mu\nu}\delta g^{\mu\nu} \tag{C4}$$

Now:

$$\delta\sqrt{-g} = \frac{\delta(-g)}{2\sqrt{-g}}$$

and recalling equation (10b), we may write:

$$\delta g = -\,g_{\mu\nu}g\,\delta g^{\mu\nu}$$

so: $$\delta(-g) = g_{\mu\nu}g\,\delta g^{\mu\nu}$$

and: $$\delta\sqrt{-g} = \frac{1}{2\sqrt{-g}}g_{\mu\nu}g\,\delta g^{\mu\nu}$$

$$= -\frac{\sqrt{-g}}{2}g_{\mu\nu}\delta g^{\mu\nu} \tag{C5}$$

Substituting equations (C4) and (C5) into equation (C3):

$$\int_\tau \left\{\sqrt{-g}\left[\delta\mathcal{L} - \left(g^{\mu\nu}\delta\mathcal{R}_{\mu\nu} + \mathcal{R}_{\mu\nu}\delta g^{\mu\nu}\right)/2\kappa\right] + \left(\mathcal{L} - \frac{\mathcal{R}}{2\kappa}\right)\left(-\frac{\sqrt{-g}}{2}g_{\mu\nu}\delta g^{\mu\nu}\right)\right\} d\tau = 0 \tag{C6}$$

In equation (29c), changing σ to ν, substituting from equation (7) and varying with respect to $\Gamma^{\beta}_{\mu\nu}$:

$$\delta\mathcal{R}_{\mu\nu} = -\frac{\partial}{\partial x^\nu}\delta\Gamma^{\beta}_{\mu\beta} + \frac{\partial}{\partial x^\beta}\delta\Gamma^{\beta}_{\mu\nu} + \Gamma^{\alpha}_{\mu\beta}\delta\Gamma^{\beta}_{\alpha\nu} + \Gamma^{\beta}_{\alpha\nu}\delta\Gamma^{\alpha}_{\mu\beta} - \Gamma^{\alpha}_{\mu\nu}\delta\Gamma^{\beta}_{\alpha\beta} - \Gamma^{\beta}_{\alpha\beta}\delta\Gamma^{\alpha}_{\mu\nu}$$

Multiplying by $g^{\mu\nu}$ and rewriting the resulting first two terms on the right:

$$g^{\mu\nu}\delta\mathcal{R}_{\mu\nu} = -\frac{\partial}{\partial x^\nu}\left(g^{\mu\nu}\delta\Gamma^\beta_{\mu\beta}\right) + \delta\Gamma^\beta_{\mu\beta}\frac{\partial g^{\mu\nu}}{\partial x^\nu} + \frac{\partial}{\partial x^\beta}\left(g^{\mu\nu}\delta\Gamma^\beta_{\mu\nu}\right) - \delta\Gamma^\beta_{\mu\nu}\frac{\partial g^{\mu\nu}}{\partial x^\beta} + g^{\mu\nu}\Gamma^\alpha_{\mu\beta}\delta\Gamma^\beta_{\alpha\nu}$$
$$+ g^{\mu\nu}\Gamma^\beta_{\alpha\nu}\delta\Gamma^\alpha_{\mu\beta} - g^{\mu\nu}\Gamma^\alpha_{\mu\nu}\delta\Gamma^\beta_{\alpha\beta} - g^{\mu\nu}\Gamma^\beta_{\alpha\beta}\delta\Gamma^\alpha_{\mu\nu} \qquad \text{(C7)}$$

Changing summation index τ to α and non-summation index σ to β, rewrite equation (16) as follows:

$$\frac{\partial g^{\mu\nu}}{\partial x^\beta} = g^{\mu\alpha}\Gamma^\nu_{\alpha\beta} + g^{\nu\alpha}\Gamma^\mu_{\alpha\beta}$$

so multiplying by $\delta\Gamma^\beta_{\mu\nu}$:

$$\delta\Gamma^\beta_{\mu\nu}\frac{\partial g^{\mu\nu}}{\partial x^\beta} = g^{\mu\alpha}\Gamma^\nu_{\alpha\beta}\delta\Gamma^\beta_{\mu\nu} + g^{\nu\alpha}\Gamma^\mu_{\alpha\beta}\delta\Gamma^\beta_{\mu\nu}$$
$$= g^{\mu\nu}\Gamma^\beta_{\nu\alpha}\delta\Gamma^\alpha_{\mu\beta} + g^{\mu\nu}\Gamma^\alpha_{\mu\beta}\delta\Gamma^\beta_{\alpha\nu} \qquad \text{(C8)}$$

after, on the right, interchanging summation indices α and ν then α and β in the first term and α and μ in the second term. Similarly:

$$\frac{\partial g^{\mu\nu}}{\partial x^\nu} = g^{\mu\alpha}\Gamma^\nu_{\alpha\nu} + g^{\nu\alpha}\Gamma^\mu_{\alpha\nu}$$

$$\delta\Gamma^\beta_{\mu\beta}\frac{\partial g^{\mu\nu}}{\partial x^\nu} = g^{\mu\alpha}\Gamma^\nu_{\alpha\nu}\delta\Gamma^\beta_{\mu\beta} + g^{\nu\alpha}\Gamma^\mu_{\alpha\nu}\delta\Gamma^\beta_{\mu\beta}$$
$$= g^{\mu\nu}\Gamma^\alpha_{\nu\alpha}\delta\Gamma^\beta_{\mu\beta} + g^{\mu\nu}\Gamma^\alpha_{\mu\nu}\delta\Gamma^\beta_{\alpha\beta} \qquad \text{(C9)}$$

after, on the right, interchanging summation indices α and ν in the first term and α and μ in the last term.

Substitute (C8) and (C9) in (C7) and simplify:

$$g^{\mu\nu}\delta\mathcal{R}_{\mu\nu} = -\frac{\partial}{\partial x^\nu}\left(g^{\mu\nu}\delta\Gamma^\beta_{\mu\beta}\right) + g^{\mu\alpha}\Gamma^\nu_{\alpha\nu}\delta\Gamma^\beta_{\mu\beta} + \frac{\partial}{\partial x^\beta}\left(g^{\mu\nu}\delta\Gamma^\beta_{\mu\nu}\right) - g^{\mu\nu}\Gamma^\beta_{\alpha\beta}\delta\Gamma^\alpha_{\mu\nu}$$

Substituting into the second term of an expanded equation (B6), we get:

$$\int_\tau \frac{g^{\mu\nu}\delta\mathcal{R}_{\mu\nu}}{2\kappa}\sqrt{-g}d\tau$$
$$= \int_\tau \frac{\sqrt{-g}}{2\kappa}\left[-\frac{\partial}{\partial x^\nu}\left(g^{\mu\nu}\delta\Gamma^\beta_{\mu\beta}\right) + g^{\mu\alpha}\Gamma^\nu_{\alpha\nu}\delta\Gamma^\beta_{\mu\beta} + \frac{\partial}{\partial x^\beta}\left(g^{\mu\nu}\delta\Gamma^\beta_{\mu\nu}\right) - g^{\mu\nu}\Gamma^\beta_{\alpha\beta}\delta\Gamma^\alpha_{\mu\nu}\right]d\tau$$

Recalling equation (17), for $\sqrt{-g} = 1$ the equation above becomes:

$$\int_\tau \frac{g^{\mu\nu}\delta\Re_{\mu\nu}}{2\kappa}\sqrt{-g}d\tau = \frac{1}{2\kappa}\int_\tau \frac{\partial}{\partial x^\beta}(g^{\mu\nu}\delta\Gamma^\beta_{\mu\nu})d\tau \tag{C10}$$

Now consider the covariant derivative of the covariant 4-vector and transforms of both denoted by the prime:

$$A_{\mu\nu} = \frac{\partial A_\mu}{\partial x^\nu} + \Gamma^\tau_{\mu\nu}A_\tau = \frac{\partial x^{\mu'}}{\partial x^\mu}\frac{\partial x^{\nu'}}{\partial x^\nu}A_{\mu'\nu'} \quad \text{and} \quad A_\mu = \frac{\partial x^{\mu'}}{\partial x^\mu}A_{\mu'}$$

So substituting:

$$A_{\mu\nu} = \frac{\partial}{\partial x^\nu}\left(\frac{\partial x^{\mu'}}{\partial x^\mu}A_{\mu'}\right) + \Gamma^\tau_{\mu\nu}\frac{\partial x^{\tau'}}{\partial x^\tau}A_{\tau'}$$

$$= \frac{\partial^2 x^{\mu'}}{\partial x^\nu \partial x^\mu}A_{\mu'} + \frac{\partial x^{\mu'}}{\partial x^\mu}\frac{\partial A_{\mu'}}{\partial x^\nu} + \Gamma^\tau_{\mu\nu}\frac{\partial x^{\tau'}}{\partial x^\tau}A_{\tau'}$$

$$= \frac{\partial^2 x^{\tau'}}{\partial x^\nu \partial x^\mu}A_{\tau'} + \frac{\partial x^{\mu'}}{\partial x^\mu}\frac{\partial A_{\mu'}}{\partial x^\nu} + \Gamma^\tau_{\mu\nu}\frac{\partial x^{\tau'}}{\partial x^\tau}A_{\tau'}$$

after change of summation index μ' to τ' in the first term on the left. So we may write:

$$A_{\mu'\nu'} = \frac{\partial x^\mu}{\partial x^{\mu'}}\frac{\partial x^\nu}{\partial x^{\nu'}}A_{\mu\nu} = \frac{\partial x^\mu}{\partial x^{\mu'}}\frac{\partial x^\nu}{\partial x^{\nu'}}\left(\frac{\partial^2 x^{\tau'}}{\partial x^\nu \partial x^\mu}A_{\tau'} + \frac{\partial x^{\mu'}}{\partial x^\mu}\frac{\partial A_{\mu'}}{\partial x^\nu} + \Gamma^\tau_{\mu\nu}\frac{\partial x^{\tau'}}{\partial x^\tau}A_{\tau'}\right)$$

Rewrite this as:

$$A_{\mu'\nu'} = \frac{\partial A_{\mu'}}{\partial x^{\nu'}} + \left(\frac{\partial x^\mu}{\partial x^{\mu'}}\frac{\partial x^\nu}{\partial x^{\nu'}}\frac{\partial^2 x^{\tau'}}{\partial x^\nu \partial x^\mu} + \frac{\partial x^\mu}{\partial x^{\mu'}}\frac{\partial x^\nu}{\partial x^{\nu'}}\frac{\partial x^{\tau'}}{\partial x^\tau}\Gamma^\tau_{\mu\nu}\right)A_{\tau'} = \frac{\partial A_{\mu'}}{\partial x^{\nu'}} + \Gamma^{\tau'}_{\mu'\nu'}A_{\tau'}$$

where

$$\Gamma^{\tau'}_{\mu'\nu'} = \frac{\partial x^\mu}{\partial x^{\mu'}}\frac{\partial x^\nu}{\partial x^{\nu'}}\frac{\partial^2 x^{\tau'}}{\partial x^\nu \partial x^\mu} + \frac{\partial x^\mu}{\partial x^{\mu'}}\frac{\partial x^\nu}{\partial x^{\nu'}}\frac{\partial x^{\tau'}}{\partial x^\tau}\Gamma^\tau_{\mu\nu}$$

So $\Gamma^\tau_{\mu\nu}$ is not a tensor due to the presence of the first term on the right. However,

$$S^{\tau'}_{\mu'\nu'} = \hat{\Gamma}^{\tau'}_{\mu'\nu'} - \Gamma^{\tau'}_{\mu'\nu'} = \frac{\partial x^\mu}{\partial x^{\mu'}}\frac{\partial x^\nu}{\partial x^{\nu'}}\frac{\partial x^{\tau'}}{\partial x^\tau}(\hat{\Gamma}^\tau_{\mu\nu} - \Gamma^\tau_{\mu\nu})$$

is a tensor as its transformation shows.

A variation is a small difference so the quantity $\delta\Gamma^\beta_{\mu\nu}$ is a tensor. Let's denote $\delta\Gamma^\beta_{\mu\nu}$ by $S^\beta_{\mu\nu}$ so equation (C10) becomes:

$$\int_\tau \frac{g^{\mu\nu}\delta\Re_{\mu\nu}}{2\kappa}\sqrt{-g}d\tau = \frac{1}{2\kappa}\int_\tau \frac{\partial}{\partial x^\beta}(g^{\mu\nu}S^\beta_{\mu\nu})d\tau = \frac{1}{2\kappa}\int_\tau \frac{\partial S^\beta}{\partial x^\beta}d\tau \tag{C11}$$

after metric contraction. For $\sqrt{-g} = 1$, equation (23b) gives $\partial S^\beta/\partial x^\beta$ as the divergence of the 4-vector S^β. The divergence theorem implies that the divergence of S^β will vanish if, on the surface of τ, $\delta\Gamma^\beta_{\mu\nu}$ vanishes. With $\delta\Gamma^\beta_{\mu\nu}$ being the variation of the gravitational field components, for a 4-volume τ of vast extents as required by equation (B1), this is conceivable. If this holds, then the left side of (C11) also vanishes. So for $\sqrt{-g} = 1$ and arbitrary $\delta g^{\mu\nu}$, equation (C6) requires:

$$\frac{\delta\mathcal{L}}{\delta g^{\mu\nu}} - \frac{R_{\mu\nu}}{2\kappa} - \frac{g_{\mu\nu}}{2}\left(\mathcal{L} - \frac{R}{2\kappa}\right) = 0$$

where, for $\sqrt{-g} = 1$, $R = \mathfrak{R}$ and $R_{\mu\nu} = \mathfrak{R}_{\mu\nu}$. The equation above may be rewritten as:

$$R_{\mu\nu} - \frac{g_{\mu\nu}}{2}R = -\kappa\left(g_{\mu\nu}\mathcal{L} - \frac{2\delta\mathcal{L}}{\delta g^{\mu\nu}}\right) \qquad \text{(C12)}$$

This becomes equation (45) – the general field equation of gravity – if we may make the following identification:

$$T_{\mu\nu} = g_{\mu\nu}\mathcal{L} - \frac{2\delta\mathcal{L}}{\delta g^{\mu\nu}}$$

In the case of the absence of matter, $\mathcal{L}$ vanishes identically and equation (C12) becomes:

$$R_{\mu\nu} - \frac{g_{\mu\nu}}{2}R = 0 \qquad : \sqrt{-g} = 1$$

which is satisfied for $R_{\mu\nu} = 0$ (as then $R = g^{\mu\nu}R_{\mu\nu} = 0$) which is the matter-free equation of gravity as given in equation (32).

Sources

[1] "The Principle of Relativity",
H. A. Lorentz, A. Einstein, H. Minkowski, and H. Weyl
Dover Publications, Inc, 1952

[2] "Reflections on Relativity"
Kevin Brown
[Online Article], cited January 11, 2011
http://www.mathpages.com/rr/rrtoc.htm

[3] "The Meaning of Relativity",
Albert Einstein
ElecBook London.
[Online Article], cited May 31, 2010
http://www.combat-diaries.co.uk/diary29/Link%2014%20Einstein.PDF

[4] "Newtonian Dynamics",
Richard Fitzpatrick
[Online Article], cited May 31, 2010
http://farside.ph.utexas.edu/teaching/336k/Newton/Newton.html

[5] "Relativity in the Global Positioning System",
Neil Ashby,
Living Rev. Relativity, 6, (2003), 1.
[Online Article]: cited May 31, 2010
http://www.livingreviews.org/lrr-2003-1

[6] "Momentum of the Pure Radiation Field"
Bo Lehnert,
Progress in Physics, Volume 1, January 2007
[Online Article]: cited July 11, 2010
http://www.ptep-online.com/index_files/2007/PP-08-04.PDF

[7] "Relativity and the Problem of Space"
Albert Einstein
[Online Article]: cited July 11, 2010
http://www.relativitybook.com/resources/Einstein_space.html

[8] "Gravity, Metrics and Coordinates"
Edmund Bertschinger,
[Online Article], cited July 11, 2010
http://web.mit.edu/edbert/Alexandria/notes2.pdf

[9] "Relativity and the Global Positioning System",
Neil Ashby,
[Online Article]: cited April 10, 2010
http://www.fing.edu.uy/if/cursos/fismod/GPS.pdf

[10] "Navstar GPS Space Segment/Navigation User Interfaces"
Interface Specification
IS-GPS-200 Revision E IS-GPS-200E
8 June 2010
Global Positioning System Wing (GPSW)
Systems Engineering & Integration
http://www.navcen.uscg.gov/pdf/gps/IS-GPS-200E_Final_08Jun10.pdf

[11] "A Brief Review of Basic GPS Orbit Interpolation Strategies"
Mark Schenewerk
GPS Solutions (2003) 6:265–267
[Online Article]: cited December 26, 2010
http://www.colorado.edu/ASEN/asen6090/Schenewerk.pdf

[12] "Guage Invariant Perturbations in Multi-component Fluid Cosmologies"
P. K. S. Dunsby
Classical Quantum Gravity 8 (1991)
[Online Article]: cited January 5, 2011
http://www.mth.uct.ac.za/~peter/pubs/cqg91.pdf

[13] "Datum Transformations of GPS Positions"
Application Note
µ-blox ag (5th July 1999)
[Online Article]: cited February 5, 2011
http://www.microem.ru/pages/u_blox/tech/dataconvert/GPS.G1-X-00006.pdf

[14] "Newton's Principia: The Mathematical Principles of Natural Philosophy "
Isaac Newton
Daniel Adee (1846)
[Online Article]: cited February 13, 2011
http://www.archive.org/stream/newtonspmathema00newtrich#page/n87/mode/2up/search/time

[15] "Lecture Notes on General Relativity"
Sean M. Carroll (December 1997)
[Online Article]: cited December 1, 2011
http://www.astrohandbook.com/ch10/general_relativity.pdf

Notes

1 *Space and Time,* H. Minkowski, p75 [1].

2 *The Foundation of the General Theory of Relativity,* A. Einstein, p117 [1].

3 [7] This clear statement of Einstein that comprises a whole paragraph seems to be forgotten.

4 p3 [8]. Bertschinger goes on to declare: "Spacetime itself is an actor."

5 "And if the meaning of words is to be determined by their use, then by the names time [and] space... their measures are properly to be understood." *The Principia,* Isaac Newton, p82 [14]

6 "For by theoretical consideration of processes which take place relatively to a system of reference with uniform acceleration, we obtain information as to the career of processes in a homogenous gravitational field." *Gravitation and Light*, A. Einstein, p101 [1]

7 Indeed, "the laws of kinematics are to be interpreted as laws which describe the relations of measuring bodies and clocks [charts]." *The Foundation of the General Theory of Relativity*, A. Einstein, p112 [1]

8 "It will be seen from these reflections that in pursuing the general theory of relativity we shall be led to a theory of gravitation, since we will be able to 'produce' a gravitational field merely by changing the system of coordinates." *The Foundation of the General Theory of Relativity*, A. Einstein, p114 [1].

9 p113 [1].

10 "But the credit of first recognizing clearly... belongs to A. Einstein. Thus time, as a concept unequivocally determined by phenomena, was first deposed from its high seat." *Space and Time,* H. Minkowski, p82 [1].

11 "for ds^2 is a quantity ascertainable by rod-clock measurement of point-events infinitely proximate in space-time, and defined independently of any particular choice of coordinates." p119 [1].

12 *The Foundation of the General Theory of Relativity*, A. Einstein, p115 [1].

13 *The Foundation of the General Theory of Relativity*, A. Einstein, p118 [1].

14 "If we then introduce, further, $\sqrt{-1}t = s$ in place of *t*, the quadratic differential expression

$$d\tau^2 = -dx^2 - dy^2 - dz^2 - ds^2$$

thus becomes perfectly symmetrical in x, y,z, s; and this symmetry is communicated to any law which does not contradict the world-postulate." *Space and Time,* H. Minkowski, p88 [1].

15 "Where ds is measured directly by a measuring-rod and $d\tau$ by a clock at rest relatively to the system: these are the naturally measured lengths and times. Since ds^2, on the other hand, is known in terms of the co-ordinates x^ν employed in finite regions, in the form

$$ds^2 = g_{\mu\nu} dx^\mu dx^\nu$$

we have the possibility of getting the relation between naturally measured lengths and times, on the one hand, and the corresponding differences of co-ordinates, on the other hand." Albert Einstein, p90 [3]. [Notation changed for conformity with text.]

16 *The Foundation of the General Theory of Relativity,* A. Einstein, p145 [1].

17 This aspect of the gravitational field may have triggered conceptualization of the 'spacetime fabric'.

18 "In the immediate neighbourhood of an observer, falling freely in a gravitational field, there exists no gravitational field. We can therefore always regard an infinitesimally small region of the space-time continuum as Galilean." p64 [3]

19 *The Foundation of the General Theory of Relativity,* A. Einstein p141 [1].

20 *The Foundation of the General Theory of Relativity,* A. Einstein p151 [1].

21 p27 [6]

22 The binary expansion is much used hereafter to attain approximations, often to a first order of $1/c^2$ which is a very small quantity.

23 *The Foundation of the General Theory of Relativity,* A. Einstein, p160 [1].

24 The following discussion follows the approach given in [2].

25 However, Einstein's historical expression of the relativistic precession as given in his *FGTR* (p163 [1]) is:

$$\frac{24\pi^3 a^2}{T^2 c^2 (1 - e^2)}$$

This may be obtained from the common expression by use of the equation of the period of an orbit:

$$T = 2\pi (a^3/GM)^{\frac{1}{2}}$$

26 p10 [5]

27 p12 [5]

28 p11 [5]

29 p12 [5]

30 p45 [9]

31 CDMA requires that the C/A code (the spreading signal) is a pseudorandom number (PRN). In GPS, the navigational data stream at 50 bps is spread by XORing it with the C/A code running at 1.023 Mbps.

32 The ECEF used in GPS is the one defined in the World Geodetic System 1984 (WGS84). This is a Cartesian coordinate system with its origin at the centre of mass of the earth, its z-axis is aligned with the IERS (International Earth Rotation and Reference Systems Service) Reference Pole also called the International Reference Pole (IRP), its x-axis is aligned with the International Reference Meridian (IRM) and the y-axis completes a right-handed orthogonal coordinate system. The IRM is 5.31 arc-seconds east of the Prime (Greenwich) Meridian.

33 The ECI frame of reference is defined conveniently as being coincident with the ECEF at a particular time. For small durations, the movement of the earth's axis of rotation in a sidereal ECI is negligible.

34 UTC is the principal civil timekeeping standard. It is based on International Atomic Time (TAI) that is a weighted average of about two hundred high-precision atomic clocks. The weighting favours the more stable clocks so the average time kept is more stable than any clock in the set.

35 For the calculation of LLA coordinates from GPS ECEF coordinates please refer to p3 [13].

36 The ECI does rotate about the sun but this influence is negligible in the context of GPS. Also, there is a negligible gravitational effect on the speed of light.

37 In this the normal case, Δt is negative. p123 [10].

38 p86 [10]

39 p265 [11]

40 p10 [5]

www.ingramcontent.com/pod-product-compliance
Lightning Source LLC
Chambersburg PA
CBHW081310170526
45166CB00011B/3471
* 9 7 8 1 4 7 5 0 1 3 5 5 9 *